# Blame It on the Weather

# Blame It on the Weather

## STRANGE WEATHER FACTS

David Phillips
Michael Parfit
Suzanne Chisholm

KEY PORTER BOOKS

**Canadian Cataloguing in Publication Data**

Phillips, D. W.
Blame it on the weather : strange weather facts / David Phillips, Michael Parfit and Suzanne Chisholm. — Rev. ed.

Includes index.
ISBN 1-55263-516-3

1. Weather—North America—Miscellanea. 2. Meteorology—Miscellanea.
I. Parfit, Michael II. Chisholm, Suzanne J. (Suzanne Judith), 1970– III. Title.

QC861.2.P44 2002 551.657 C2002-902869-8

The publisher gratefully acknowledges the support of the Canada Council for the Arts and the Ontario Arts Council for its publishing program.

We acknowledge the financial support of the Government of Canada through the Book Publishing Industry Development Program for our publishing activities.

Key Porter Books Limited
70 The Esplanade
Toronto, Ontario
Canada M5E 1R2

www.keyporter.com

Design: Peter Maher
Electronic formatting: Jean Lightfoot Peters

Printed and bound in Canada

02 03 04 05 06 6 5 4 3 2 1

# Contents

# Weatherwise

# Living with Pinatubo's Fallout

Since the fiery eruption of Mount Pinatubo in the Philippines in June 1991, the weather in many parts of the world has taken an unusual turn. Global temperatures in 1992 and 1993 fell below the record warm levels of the 1980s—a phenomenon many climatologists attribute to the large amounts of gases and ash that Pinatubo propelled into the atmosphere.

After 635 years of dormancy, Pinatubo erupted with a blast equated to 2,500 Hiroshima bombs. The volcano spewed a plume of gray ash, fiery debris, and hot gases 24 miles high—the largest of any eruption this century. Within five weeks, winds about 12 miles above the surface had spread a veil of volcanic dust around some 50 percent of the planet. Veteran astronauts aboard the space shuttle *Atlantis* that summer reported that they had never seen the earth look so hazy.

**The death toll from Mount St. Helens' eruption in 1980: 57 people, 5,000 black-tailed deer, 200 black bears, 1,500 elk, and countless birds.**

Following the eruption, satellite and aircraft observations revealed a 25 to 30 percent decline in the solar radiation reaching the ground. Temperatures began dropping soon after Mount Pinatubo erupted. In 1992, Pinatubo's shadowy plume cut almost a degree from the average global temperature of about 59°F, nearly counteracting, at least temporarily, 100 years of global greenhouse warming. The cooling might have been even greater had the eruption not coincided with an El Niño—a pronounced warming of the Pacific Ocean water off the coast of South America known to cause a global warming of 0.36°F.

A similar coincidence of a low-latitude volcanic eruption and an El Niño warming occurred in 1982, when El Chichón erupted in Mexico. Rich in sulfur dioxide, El Chichón produced a dust veil 20 times larger than that from Washington State's Mount St. Helens on May 18, 1980,

Mount St. Helens erupted on March 30, 1980. *Canapress*

enough to lower the surface temperature in the Northern Hemisphere by a few tenths of a degree. Precisely when and how much cooling took place is impossible to determine, however, because the concurrent El Niño warming was the largest such event this century.

The notion that volcanic eruptions can influence the weather dates back more than 200 years. The ever-curious Benjamin Franklin was among the first to propose the connection. While serving as American ambassador to France in 1785, Franklin speculated that Europe's unusually harsh winter in 1783–84 was due in part to a "permanent dry fog" that had drifted eastward following the 1783 explosion of the Laki volcano in Iceland.

**The eruption of Krakatoa on a small island between Java and Sumatra on August 27, 1883, registered on barometers halfway around the world in London. In fact, it was still registering nine days later, when the echoes of its volcanic report were making their seventh circuit of the Earth.**

Today it is generally accepted that large volcanic eruptions can affect the weather. The extent of that influence

**Mary Shelley's *Frankenstein* is instantly recognized as one of the world's great horror stories, but few people know that it owes its origins to the weather. The book was written in the summer of 1816, the famous "year without a summer," when unseasonably cold weather struck parts of North America and Europe. Fourteen months earlier, volcano Tamboro had erupted, spreading a veil of dust that altered weather patterns around the world. The rainy, chilly summer on Lake Geneva in Switzerland prevented Mary Shelley and her companions from going outdoors. To pass the time, they began to tell ghost stories. A ghost-story-writing contest developed and Mary Shelley wrote *Frankenstein*.**

depends on the composition and height of the plume and on the latitude of the volcano.

The plume consists of ash and gases, mainly water vapor, carbon dioxide, and sulfur dioxide. It is the volume of sulfur dioxide expelled into the stratosphere, not the more spectacular clouds of dust and ash, that most affects global weather. The more emitted, the greater the impact will likely be. Within a few months of an eruption, gravity draws most of the ash particles into the lower atmosphere, where rain and snow remove them within a few days or weeks. The sulfur dioxide in the stratosphere bonds rapidly with water vapor in the presence of sunlight, producing tiny droplets of sulfuric acid that coalesce into a dense haze of aerosols—fine liquid or solid particles. These acidic droplets are so small and light that they remain in the stratosphere for one to three years, deflecting some of the sun's energy and cooling the earth's lower atmosphere and surface. This cooling is partly offset by warming, since the aerosol also absorbs infrared radiation reflected from the earth's surface. In the case of Pinatubo, the overall result was a slight cooling of the lower atmosphere.

The height of the plume helps to determine how widespread and persistent the effects will be. The higher it rises, the longer volcanic gases will stay in the atmosphere and the longer they are likely to affect the weather. For example, the plume from the eruption of Mount St. Helens reached a height of only 15 miles—considerably lower than

Pinatubo's—and contained little sulfur dioxide. The most notable weather effects were cooler temperatures near the mountain for three or four days immediately following the eruption, as volcanic ash temporarily blocked out the sun. But no long-term or widespread weather effects were recorded.

**Rafts of floating pumice, some of them thick enough to support men, crossed the Indian Ocean in 10 months, and were still seen two years after the eruption of Krakatoa.**

As for the latitude of the volcano, scientists suspect that large sulfur-rich eruptions near the equator have the greatest impact on weather. Air-circulation patterns there are more likely to carry the volcanic debris into both hemispheres, clouding more of the planet. Plumes from mid-latitude eruptions disperse well east and west, but are much slower to spread north or south. In terms of global effect, Mount St. Helens' eruption was a non-event. Estimated at one-tenth the energy of Krakatoa's, Mount St. Helens was too far north, erupted sideways (not straight up) and threw too little of the sulfur-rich debris needed to significantly affect the weather. Furthermore, the ejecta's height of only about 15 miles was relatively low as such energetic eruptions go.

**For months after the eruption of Krakatoa, the fine volcanic dust and aerosols caused such brilliant red sunset afterglows that on several occasions fire engines were called out in northeastern United States to extinguish apparent fires. Vivid sunsets, white suns at noon, and rings of pink, red, orange-rose, and brown continued to be seen for the next three years.**

Cooling aside, there is no proven link between increased volcanic activity and unusual weather. Major volcanic eruptions are often followed by spells of gloomy weather, cool and wet summers, prolonged winters, droughts, and early frosts. However, the same conditions could also be attributed to weather's random variability.

Looking back at some of the big eruptions in the past two centuries, we can easily see how people started making a connection between volcanoes and bad weather. The April 1815 eruption of Tamboro in Indonesia—considered the most explosive volcanic eruption in the past 5,000 years—

emitted about 16 times more ash and gas than Mount Pinatubo and killed 92,000 people. A year later, Tamboro's cloud caused remarkably strange weather across Europe and northeastern North America. With average surface temperatures 2°F to more than 5°F below normal, heavy snows fell in June and hard frosts occurred in July and August, killing crops and denuding trees. There was a summer blizzard in Connecticut, snow and sleet in Vermont, and flurries in Massachusetts. Many birds and newly shorn sheep died from exposure, and crop failures in New England brought ruin, famine, and disease. As a result, 1816 became widely known as "the year without a summer."

**The famous climate researcher H. H. Lamb found that most of the 20th century was markedly quiet in the number of significant volcanic explosions compared with the period from 1500 to 1912. The only notable eruptions since 1912 have been Mount Agung in 1963, El Chichón in 1982, and Pinatubo in 1991. By contrast, during the previous four centuries, there was, on average, one eruption every four years.**

Similar weather has been associated with other explosive volcanoes. Krakatoa—on an island in Indonesia between Java and Sumatra—erupted in May and August 1883, propelling clouds of ash and gas 30 miles high. The explosions were heard 3,000 miles away in Australia, and the sea waves generated by the eruptions were detected in the Atlantic Ocean 10,000 miles away. The region was plunged into darkness for almost three days. In addition, Northern Hemisphere temperatures cooled about 1°F to 1.5°F, and blazing red sunsets and strange haloes around the sun and moon were created for a couple of years following the eruption. On November 27, 1883, fire engines were called out in New York City because of the red afterglow of a sunset.

Since the 1980 eruption of Mount St. Helens, scientists have been using satellites and airplanes to learn more about how volcanoes affect the weather. It now appears that the most likely short-lived regional effect is a modest temperature reduction of typically 0.5°F to 1°F in the year following an eruption and a 0.2°F to 0.5°F decline two years after. In theory, this cooling and increased cloud cover could in turn change the normal pattern of weather circulation for several months, including the path of the jet stream and the presence or absence of stationary high-pressure systems that are often associated with weather extremes and spells of unusual weather. During the summer of 1992—a year after Mount Pinatubo erupted—the jet stream wandered farther south than normal, bringing more rain, cooler temperatures, and increased cloud cover to central and eastern Canada. The daily weather map in July looked more like one in May.

**In addition to cooling the Earth, Pinatubo diminished the Earth's protective ozone layer as it came into contact with the ash and gas. Some experts estimate that as much as 15 percent of the ozone was destroyed, though ozone does replace itself over time.**

Atmospheric scientists believe that in addition to cooling the Earth, Mount Pinatubo's gases, some 25 million tons, may have contributed to the destruction of as much as 15 percent of the Earth's imperiled ozone layer in the Northern Hemisphere. With less ozone, more harmful ultraviolet radiation reaches the surface. Fortunately, ozone can slowly rebuild itself. Pinatubo's blast also produced many colorful and enduring sunrises and sunsets, striking solar rings, green suns, and blue moons.

Because a single volcanic eruption affects the weather for only a couple of years, some scientists question whether volcanoes can actually change the climate in the long term. For many years, volcanoes were considered a prime cause for initiating glaciers. Some periods of global cooling and increased glaciation have coincided with exceptionally high volcanic activity. For example, the period from 1500 to 1850, known as the Little Ice Age, experienced several major volcanic eruptions and some of the worst summers and coolest winters on record across North America and Europe. However, there is little solid evidence to prove a link between volcano activity and ice ages. Indeed, a case can be made for the reverse. The stress of continental glaciers pressing down on the Earth's crust may trigger volcanism.

# Arctic Haze

Crystal-clear air, pure and sparkling, used to be an Arctic hallmark. Travelers in the 19th century reported almost unbelievable visibility, spotting mountain peaks 120 miles away, poking over the horizon. Today, between February and May, you'd be lucky to see 20 miles through the hazy Arctic atmosphere. The Arctic's winter air has become tarnished, dimmed by a dirty blanket of reddish brown smog that clogs the air, acidifies the precipitation, and leaves telltale smears on the landscape. Known simply as Arctic haze, it arrives each fall and winter—a totally unexpected phenomenon recorded nowhere else on earth.

**It has been estimated that Arctic air pollution increased by 75 percent between 1956 and 1986, paralleling a doubling of acidic sulfur-dioxide emissions in Europe and the former Soviet Union.**

People puzzled over the Arctic's hazy air long before they knew its origin. A century ago, Norwegian adventurers trekking across the polar ice pondered mysterious dark stains on the snow surface and in melt pools. As far back as 1914, an Inuit guide referred to the misty haze that shrouded distant summits as *poo-jok*, which means "fog" in Inuktitut, the Inuit language. An early theory held that the haze was sand blowing in from the Gobi Desert. A few scientists argued that it was just ice crystals and blowing snow. Pilots like Royal Canadian Air Force Wing Commander Keith Greenaway flying weather reconnaissance over the Canadian northwest in the late 1940s were surprised by the thickness of this "ice crystal haze." Others suggested northern sources such as smoke from the naturally burning Smoking Hills near the Mackenzie River delta or exhaust from machines involved in oil exploration on the Beaufort Sea shelf. What they had to rule out, though, were nearby towns, mines, and oil and gas refineries. The Arctic just doesn't have major pollution sources.

**Theoretically, when the air is clean, one can see up to 120 miles. However, haze in the Arctic can limit visibility to a mere 20 miles or less.**

So entrenched was the myth of the pristine North that scientists refused to believe the Arctic could be widely polluted. "It was not until the 1970s that a series of air-chemistry measurements conducted first at Barrow, Alaska, and then in Canada showed that Arctic haze is not natural but more human in origin," says a research scientist.

**So far, at least, the air at the South Pole remains pure—perhaps the cleanest air on Earth and about 10 times cleaner than that found in the Arctic. The Southern Hemisphere contains relatively few sources of industrial pollution, and its wind patterns do not sweep pollution towards the pole.**

The haze, in fact, is an airborne swamp of industrial pollutants, mostly sulfur and nitrogen compounds found as a gas, together with aerosols. Over time these transform into micro-particles of sulfuric and nitric acids, similar to the acid-rain pollution that plagues Europe and North America.

Using aircraft equipped with lasers and high-volume vacuum samplers, scientists have flown back and forth through the haze layers in order to measure its extent and concentration. The haze is Arctic-wide, blanketing virtually the entire region north of 60°N. It makes the Arctic winter air 10 to 20 times more polluted than Antarctica's and some 10 times more polluted than that over the least polluted non-industrial regions elsewhere in North America.

**The most northerly permanent environmental laboratory in the world is found at Alert in Canada's Northwest Territories. The laboratory was opened in August 1986. Its purpose is to measure the concentration of industrial pollutants such as carbon dioxide, halogenated organic compounds, and sulfates, and to monitor the changes in their concentrations over time.**

Where does the stuff come from? How does it reach the Arctic? Why does it peak in the winter and spring and vanish in summer? Over the past 20 years, a small, highly motivated group of glaciologists, meteorologists, and atmospheric chemists from the United States, Canada, and Norway have tried to answer these questions.

To find the source of the polluted air, they first hunted for its path by

**According to French scientists, snow that fell on Greenland in 1967 contained seven times more lead than snow that fell in 1989. Concentrations of two other heavy metal pollutants, cadmium and zinc, more than halved in the same period. The decline mirrors a drop in heavy-metal pollution in the lower atmosphere in the past two decades, and a switch to unleaded gasoline.**

looking at routinely observed winds and pressure fields. But that approach was frustrated by the scarcity of meteorological data from the Arctic, and the enormous distances, some 4,800 to 6,000 miles, back to sources in mid-latitude.

The answer lay in finding the villain's chemical "fingerprints." In 1980, Len Barrie of Canada and Kenneth Rahn, a research professor at the University of Rhode Island, developed a tracing method that identified a distinctive chemical signature in an air mass, which they could trace back to its region, if not very factory, of origin. They were able to determine the type of raw fuel and ore used in refineries by comparing the isotopes in a polluted air sample found in the Arctic with those in the raw material (ores and fuel). The composition of lead in emissions from refineries in Russia, Western Europe, and North America varies, they found, depending upon the region.

Using this model, the scientists discovered that up to two-thirds of the Arctic's haze originates in the heavily industrialized nations of Eastern Europe and Russia, and the remainder in Western Europe. North America contributes less than 4 percent. This is no reason for us to feel smug: it just happens that the prevailing winds carry our pollution east over the Atlantic Ocean, where storms clean the air. However, European and Russian air masses are propelled northwest into the Arctic in winter.

In the fall and winter, the Arctic atmosphere loads up with airborne contaminants. The grimy winter air is routinely 20 to 40 times dirtier than in summer for at least three reasons related to meteorology. The great dome of colder, heavier surface air that traps pollution over the Arctic is largely a winter phenomenon. In the inversion, increased stability and stagnation form an invisible barrier 400 to 800 yards above ground. This prevents incoming contaminants from mixing vertically, and traps what little pollution is released locally from Arctic towns and power plants.

Second, the large weather systems that control the movement of pollutants into, through, and out of the Arctic are particularly vigorous in winter, strengthened by marked north-south temperature contrasts. In addition, a zone of strong northward flow is created by intense storms linked with huge continental high-pressure areas. In summer, by contrast, the weather circulation is much weaker, and air generally moves from north to south.

Finally, during the fall and winter seasons, haze particles have a longer residence time in the atmosphere than in the summer. Because the air passes over what is essentially a frozen desert, there is little rain or snow to wash out pollutants. In the summer, more abundant rains and low-lying clouds associated with drizzle and fog scrub out the pollution.

**Although Arctic winter-spring pollution levels are not as bad as those found in many US cities, they are about one-fifth the average levels around Lake Erie between Detroit, Cleveland, Buffalo, and Toronto. The Arctic concentrations are about 10 times higher than background levels in remote areas of the Northern Hemisphere.**

How long has haze been polluting the Arctic? The answer was found in the ice of Ellesmere Island by Dr. Roy Koerner, a glaciologist with the Geological Survey of Canada. In the 1980s, he pulled up drill cores from centuries of accumulated snow and ice, analyzed their acidity levels, and found a century-long record of industrial pollution. Levels of acidity in the century's first half were constant but changed dramatically after 1956, increasing by some 75 percent over the next two and a half decades. This increase mirrored the doubling of acidic sulfur-dioxide emissions worldwide and included a new mess of airborne toxic contaminants, including herbicides and pesticides such as lindane and DDT; heavy metals such as lead, mercury, and vanadium; and industrial organic compounds like solvents, dioxin, and PCBs.

Whereas we now know a fair amount about the composition, variation, and origin of the haze, we know a lot less about its effects on the precarious Arctic ecosystem and the global environment. Most obvious is the sharp reduction of visibility. But more significantly, haze alters the balance between heat and cold—and possibly, the climate. The black, sooty aerosols in the air absorb the sun's heat, warming the lower atmosphere. Soot on snow and ice also absorbs heat from the sun's rays that would otherwise be largely reflected back into the sky. As the snow and ice cover melts, less heat is reflected, which accelerates the overall warming and melts more snow and ice, and so the cycle continues.

Recently, however, there's been some good news. Scientists have learned that the trend in Arctic haze, which began accelerating in the 1950s, has apparently been stopped. And more good news: although the sulfate content has remained roughly the same, concentrations of some toxic organic contaminants and heavy metals have fallen dramatically. Levels of the insecticide lindane at one research station have dropped 90 percent since 1979, and lead concentrations have shrunk by 55 percent since 1980. This decline is partly due to international efforts to eliminate lead in gasoline and to control toxic releases.

Measures aimed at curbing pollutants that compose haze are working. Still, much work remains to be done in curtailing pollution and identifying polluting sources and pathways before the Arctic air is returned to its once-pristine condition.

# Halos, Sundogs, and Sun Pillars

One day in the summer of 1865, tragedy befell a party of British and French mountain climbers in the Alps: after making their first ascent of the Matterhorn, four men tumbled from the steep slopes and were killed. Shortly afterwards, a surviving member of the team, Edward Whymper, saw a startling and poignant phenomenon in the sky: a large circle of light with three ghostly crosses, "a strange and awesome sight... at such a moment," as Whymper wrote.

The alpinist was not gazing at a spectral vision but at an unusual combination of figures created by distorted sunlight—a halo, circle, and pillar. Its timing was astonishing but purely coincidental, and is perhaps the most famous example of such solar displays.

**The sky is often blue on Earth, but skies are not blue on every planet. The sky on Mars is pink; it's black on the moon and yellow on Venus.**

Almost everyone is familiar with lightning, rainbows, and the northern lights, but few people are aware of the dazzling variety of bright, sometimes colored spots, patches, rings, and arcs occasionally on display in the sky. Although halos, sundogs, and sun pillars occur quite often, these optics are rarely seen by the casual observer. Few people know where and when to look for them. This is a pity because these phenomena are not only beautiful, but can also tell us something about the clouds overhead and help us predict changes in the weather.

Optical phenomena are the result of light refracting (bending) through or reflecting (bouncing) off tiny, floating ice crystals in the air or high clouds. They come in a great assortment of patterns, which are determined by three factors: the shape of the crystals themselves, the path light takes through them, and the alignment of the crystals as they fall or are suspended in the air. First, although all ice crystals are six sided, like snowflakes, they have several different shapes. The most common are six-sided plates and columns, but bullet-shaped crystals, some with flat caps, also occur.

**In AD 40, the Roman College of Soothsayers foretold good fortune from an observation of three suns that appeared in the sky, i.e., sundogs. However, the soothsayers missed the mark, as the following year was not a good one for Rome.**

Second, since ice crystals are not round like raindrops, light can travel in many pathways through them, entering an end or side and exiting from a different facet; or light can reflect off the crystal surfaces. The different paths create different light patterns.

Third, the ways in which ice crystals twist and turn when falling through the air increase the variety of effects they create. Some crystals are randomly aligned in the air, while others are ordered identically, like soldiers on parade facing the same direction.

**Although sky effects are caused mostly by ice crystals, they can occur throughout the year, even on the warmest summer day, because the upper extent of the atmosphere is always extremely cold.**

Each orientation can produce its own distinct optical effect. When ice crystals are very small, constant bombardment by moving air molecules keeps them suspended randomly (at all angles to the ground). But large crystals may fall face downward, much like a dinner plate on being dropped. Each orientation produces its own distinct optical effect.

The most spectacular and commonly observed figure is the halo—a circle of light around the sun or moon. Most halos are created by light entering one of the side faces of each ice crystal and exiting through a different side face after changing direction (refracting) about 22°. The crystals must be of uniform size and point in all directions to produce the circular diffusion. Larger, fainter halos sometimes occur farther from the sun or moon, and are called "great," or 46°, halos. They are formed either when light enters the top of each crystal, refracts, and emerges from one of the sides, or when light enters one side and passes out the bottom.

**On a few clear days, you may glimpse a lone streak of emerald green light shooting straight up from the sun at the horizon. Called a green flash, it only lasts a second. The green flash can occur at sunrise or sunset, when the clear atmosphere disperses that very first or final ray of the sun's white light into the colors of the spectrum. The rays of the shorter blue end of the spectrum are scattered and lost, so the next hue, green, flashes before your eyes.**

A halo may appear as a complete ring when the sky is covered with a uniform sheet of wispy, high cirrostratus cloud. More often, however, the cloud covers only part of the sky, and as a result, only segments of a halo are visible. Usually the ring of light appears whitish, but on occasion the colors of the rainbow may be seen faintly, with red on the inside, yellow and green in the middle, and bluish white on the outside.

Solar halos are more common than lunar halos, which are seldom noticed unless the moon is full. (Because the extremely bright light of the sun can damage your eyes when looking into it, always block it out with an open hand or a book held at arm's length.) Halos occur most frequently when the sun is near the horizon and blurred by a thin veil of cirrus clouds. They can occur at any time of year, but spring is often the best viewing season.

**The best viewing weather for spotting halos, sundogs, and sun pillars includes: a cold Arctic air mass with an air temperature near the surface of about 1.5°F; bright sunshine, like that which occurs after the passage of a cold front; blowing snow or a trace of cumulus fractus clouds; moderate to strong winds.**

You may have observed bright, sometimes rainbow-colored blazes on either or both sides of the sun. These are commonly called sundogs (presumably because they "dog" or mock the sun), but the technical name for them is parhelia. They occasionally occur, less brightly, around the moon, and are called moondogs or paraselenae.

Sundogs are created when sunlight hits plate-shaped ice crystals falling in a uniform alignment with their six sides vertical. Such crystals, at least the larger ones, tend to float like parachutes with their top and bottom faces oriented horizontally. Sunlight enters one of the vertical side faces, refracts once, traverses the interior of the crystal, and refracts again as it exits. This bends the light about 22°, creating mirror solar images about 22° from the sun—slightly outside the 22° halo that often occurs at the same time. The most brilliant ones appear when cold air is dense with plate crystals, when the crystals fall without excessive flutter or tumbling, and when the sun is low in the sky. Look for a cold sunny morning or evening, when the sun is near the horizon.

The simplest interaction of sunlight or moonlight with ice crystals occurs when light reflects off the crystals to produce pillars. They are vertical streaks of light stretching to a height of 5° to 10° above, and occasionally below, the sun. Most common at sunset or sunrise, these glittering columns are created when a beam of sunlight reflects off the flat bottoms of descending plate crystals in cirrus clouds or extremely cold air. Because they are caused by reflection, sun pillars are the same color as the sun. If the sun is very low, the pillar may reflect the red hues of sunset (or sunrise). Pillars are sometimes confused with crepuscular (twilight) rays, which are rays of sunlight—often spectacular—that pierce gaps in a layer of low cloud, or radiate upward through gaps in the lumpy tops of cumuliform clouds just above or below the horizon.

Sundogs, and all related halo phenomena, were regarded with awe by early peoples and were thought to forewarn of troubled times. They were also seen as harbingers of foul weather. This is apparent in the following folk sayings:

- Ring around the moon, rain is coming soon
- When the sun is in his house, it will rain soon
- The bigger and brighter the ring, the nearer the wet
- The moon with a circle brings water in her beak
- A bright circle round the sun denotes a storm and cooler weather
- The circle of the moon never filled a pond; the circle of the sun wets a shepherd

In fact, a halo around the sun or moon is one of the better weather predictors visible in the sky. The high cirrus stratus clouds—the so-called mare's tails—that cause halos are often the first visible sign of an approaching warm front, bringing wet weather. Studies have shown that on two out of three occasions, rain or snow will arrive within 18 to 36 hours after a halo has been spotted.

**A weather observer near Bristol, England, diligently watched for halos in the sky during a 12-year period. He found that of the 80 halos sighted, 71 were around the sun and nine around the moon; 39 halos lasted less than five minutes, 11 lasted more than an hour. As for the adage that halos are a sure sign of rain or snow, he found that on only 45 occasions out of 80 did rain follow within 48 hours.**

The chances of accurately predicting rain or snow from halos depends upon how close you are to one of the corridors taken most commonly by storms. In Dallas, Texas, a halo may mean there is a 35 percent chance of rain, whereas rings over Columbus, Ohio, might indicate a 65 percent chance of rain. Columbus is in a storm track, Dallas is not. If, in addition to seeing a halo, you observe falling barometric pressure, an increasing southerly wind, and progressive thickening of cloud from the west, you can be more confident that rain or snow is on the way.

There are many other less commonly known optical effects produced by ice crystals—double halos, the Parry arc, and subsuns, to name a few. Most are not as rare as you might expect. What is rare, however, and sure to stir the imagination, is seeing two or more optical effects together, as Edward Whymper did on that fateful day in the Alps in 1865. The skies are constantly changing, so look up and treat yourself to the many spectacles that have struck awe in skywatchers throughout history.

# Snowflakes, Snowballs, and Snowrollers

## Is It Ever Too Cold to Snow?

You may think so! On the coldest days of the year, the sky is often sunny and cloudless and there is "no snow" in the air.

You may have noticed that the heaviest snowfalls with the largest and stickiest snowflakes typically occur when the temperature is near freezing rather than very low. For some Midwestern cities, March is, on average, both the mildest and snowiest of the winter months. This is because warm air holds more moisture than cold air, which means that more water can fall as snow. For instance, at −22°F, saturated air contains only one-tenth of the water vapor it would contain at +23°F. In addition, at very low temperatures, ice crystals are dry and do not bond well. Instead, they settle out as tiny single crystals rather than aggregate as snowflakes.

**The biggest snowflake recorded in the US fell near Fort Keogh, Montana. Flakes the size of small pizzas fell there on January 28, 1887.**

On the other hand, even in the coldest regions of Alaska—where there is very little water vapor in the air—very fine snow still falls. At Bettles,

Alaska, one of the coldest places in the US, an average of 12.4 inches falls each January, despite an average January temperature of −12.7°F.

So while the notion that it can be too cold to snow is erroneous, the amount of snowfall is usually less the colder it gets.

## What Are Snowrollers?

Snowrollers are giant natural snowballs. Snowrollers have a certain mystery to them, like other weather oddities such as ball lightning, white tornadoes, and waterspouts. Although they occur in the US each year, they are rare and form mostly at night or in the early morning. Moreover, they are the stuff of kids' pranks: a field of huge snowballs without telling footprints or a single eyewitness.

**Snowrollers have been observed in many parts of the world. On March 4, 1980, snowrollers formed on the island of Cyprus. The snow happened to be of the required consistency for snowrollers to form and roll to the bottom of a hill. All were subsequently found to have pebbles at their centers.**

A really good snowroller is cylindrical with hollowed-out ends, often with a hole extending through it lengthwise. As a result, snowrollers have been compared to ladies' muffs, rolls of light cotton batting, or a rolled-up carpet. They range in size from eggs to small barrels, but are typically the size of a football. Tracks behind them mark the path along which the wind has been blowing and are typically less than a half inch deep and several yards long, though they may extend the length of a football field. Snowrollers may weigh up to three pounds, but, unlike handmade snowballs, they are composed of such loose and fluffy material that they fall to pieces at the slightest handling.

Snowrollers are rare because conditions for their formation must be perfect. They need:

- a layer of newly fallen, light, feathery snowflakes, followed by
- a warming, say two or three degrees above freezing, sufficient to melt but not thaw the layers of snow, so that the soft and damp snow crystals cling to one another (which leads to rolling rather than drifting), and

- strong, gusty winds of at least 25 mph to pick up the snow and start it somersaulting along.

**Before the automobile age, the term "snowroller" was applied to horse-drawn vehicles consisting of two large wooden rollers that flattened snow on public roads to make a smooth path for sleighs.**

It also helps to have a gently sloping smooth surface, where the force of gravity helps the wind roll the snow along. The best surface is a smooth crust of old snow with no tufts of grass or other projections to impede the roller.

Sticky snow, warm air, and strong winds provide perfect snowroller conditions. However, something has to be missing: if not, snowrollers would be as common as rainbows and sundogs. What starts the snowball rolling?

Some have suggested it is a strong downward gust that scoops up fragments of snow or ice off the surface and sets the roll in motion, much like a breeze lifts the edge of a piece of paper lying on the ground. Others suggest that flakes hit the snow surface at a sharp angle, tip over, and are immediately caught by a wind gust and pushed along.

**During the Blizzard of '96, from January 6 to January 8, 100 million tons of snow fell on New York City.**

Once in motion, rollers tumble onward until the wind slackens or until they grow so large that they become lopsided and fall over, or until the ground levels off too much for the wind to propel the mass of snow farther.

Snowrollers develop somewhere in the US each winter. They are most common in Montana, Wyoming, and Colorado, along the Rocky Mountain front, where winds tend to be strong and temperatures rise rapidly. But chances are you will see snowrollers on a steep roof before you spot them in an open field.

## Are Any Two Snowflakes Alike?

It has long been said that like human fingerprints, no two snowflakes are exactly alike. Is this true? Well, it depends on how you define "alike." If alike means that snowflakes have to look the same, then the cherished belief isn't true. After all, to most observers, a snowflake is a snowflake. And there are lots of them. Over the past five billion years—the lifetime

A cylindrical snowball that formed during the night in February 1997. *Environment Canada*

of the Earth—an estimated one undecillion flakes (1 followed by 35 zeros) have blanketed the Earth. Since all snowflakes start the same way and are usually composed of six-sided or six-pointed crystals of ice, it would seem reasonable that a duplicate would one day appear, if only someone would take the time to look.

Wilson Bentley, or "Snowflake" as he was better known, of Jericho, Vermont, spent his lifetime looking for twin flakes. The farmer and amateur meteorologist put nearly 6,000 flakes under the microscope, photographing and recording each delicate feature. Eventually, he identified 100 different crystal structures, but identical snowflakes eluded him. He died in 1931, his search unresolved.

**Keeping roads in the US free of ice and snow costs some $2 billion each year.**

Others followed in his footsteps. In 1986, two researchers at the United States National Center for Atmospheric Research in Boulder, Colorado, gained some notoriety with their claim of a matching set of snow crystals collected, by means of fixed probes extending from the front of an aircraft, 7,000 yards above Wisconsin. The tiny, column-like crystals with vase-shaped hollow centers were very much alike but not structurally identical.

So why the difficulty tracking down twin flakes? The answer is that probably none exist. So many variables are at play when a snowflake develops, meteorologists say, that the odds of two following identical paths are ridiculously small.

**Snow has fallen in every state in the US, including Hawaii.**

Cumulus clouds stacked high on the horizon forewarn of a winter's storm. Along the cloud's underbelly, water droplets collect; in its upper reaches, where temperatures are well below freezing, ice crystals form. Every snowflake begins as a single ice crystal, with most forming around a solid particle known as a freezing nucleus—a floating piece of clay, fine dust, tiny pollen, sea salt, volcanic ash, or even a fleck of pollution from a car's exhaust. Water vapor in the surrounding air uses the nucleus as a platform. In a process known as sublimation, the vapor condenses into ice on the surface of this host nucleus without first going through the liquid phase. If the ice crystal continues to grow by sublimation, an individual, intricate, and visible snow crystal soon develops. Other ice crystals form without a freezing nucleus. In this case, tiny beads of supercooled water freeze directly into ice at very low temperatures. (Left alone, supercooled water droplets stay unfrozen at temperatures as low as −40°F before hardening spontaneously into ice.)

**Fresh snow falls each year on roughly a quarter of the Earth's surface, as well as about 10 percent of the oceans. Only one-third of the people in the world have ever kicked the white stuff.**

But in either situation, neither the supercooled liquid droplets nor the individual ice crystals are large enough to fall from clouds as precipitation. To become snowflakes, the ice crystals must grow many times larger; they do so mainly by sublimation of water vapor.

An ice crystal grows by attracting and absorbing water vapor molecules that build on one another in a symmetrical way, thus continually adding to its weight and size. But since each of the thousands of billions of water molecules has an astronomically large number of choices of where to attach itself to the crystal, it is very unlikely that every molecule would line up in exactly the same position as every molecule in another crystal.

Still, the ultimate shape and size of the snow crystal, hence the snowflake, depend on temperature and, to some extent, on the amount of water vapor available in the cloud. Even slight changes in the cloud environment can have a marked effect on the crystal structure:

- Crystals that grow in air at −4°F or colder, and with little available moisture, form hollow prisms or hexagonal columns similar in shape to a length of pencil
- Between −9.4°F and −0.4°F, crystals are compacted into flat wafers called hexagonal plates
- With more moisture available between 1.4°F and 10.4°F, larger crystals grow into delicate, six-pointed stars called dendrites, their shape reminiscent of human nerve cells by the same name
- In warmer air but still below freezing, crystals grow into splinter-shaped bits of ice known as needles; and at freezing, crystals take on the shape of thin hexagons

Only when the ice crystal has attracted sufficient mass does it begin its journey earthward. At some point, the ice crystal will become heavy enough to fall at about 20 inches per second. During its descent, the falling ice crystal collides and coalesces with supercooled droplets and other ice crystals. As this multi-crystal aggregate becomes larger, it falls faster. Eventually, it drops out of the cloud as a snowflake.

**In November 1997, the owner of a Canadian ski hill was commissioned to send 50 tons of fresh snow to sunny Puerto Rico. The snow made the 2,000-mile trip in a refrigerated container, and arrived in time for a four-day celebration to kick off the Christmas season.**

Nearly anything can happen to snowflakes as they drift and tumble earthward. Pieces break off, evaporate, or melt. Flakes bump into one another and sometimes bind together. Others pick up frozen water droplets and tiny particles—all processes that change a snowflake's shape, design, and water content. When falling snowflakes begin to melt, they stick to just about anything, even to each other. In some cases, hundreds and thousands of ice crystals may cling together to form gigantic snowflakes. Also, the greater the distance snowflakes fall, the larger they become. In general, snowflakes are less than a quarter of an inch in diameter. But on rare occasions, they can coalesce into gigantic snowflakes. Back on January 28, 1887, at Fort Keogh, Montana, monster snowflakes "larger than milk pans" and measuring 15 inches in diameter were reported.

For two crystals to look the same under a microscope, they would have to attract the same number of molecules of water vapor, the molecules would have to arrange themselves in the same way, and each crystal would have had to collide with the same number of other crystals during its long fall to the ground. Even the slightest change in a host of factors, from a rise in air temperature to a whiff of wind, will make two flakes different. So are there any two snowflakes that are exactly alike? Very unlikely indeed!

## Why and When We See Rainbows

Many people consider the rainbow to be the most magnificent of all sky phenomena. Its appearance has delighted children and inspired artists, poets, and composers throughout the ages. In many cultures, notably the Irish, the rainbow was believed to be a magical pathway to good fortune and happy times. Although the ancient Greeks saw it as a sign of war and death, several cultures revered it as a symbol of renewed hope. In the Bible, the rainbow is a reminder that God will never again flood the Earth. Indeed, a rainbow is often a sign that the rain has ended.

**Rainbows usually have only a fleeting existence, but on August 14, 1979, a rainbow was seen on the coast of North Wales that lasted for three hours—perhaps a world record.**

A rainbow is merely sunlight. There is nothing material about it; it is an optical illusion like a mirage or a halo. To see a rainbow the proper angle must be set up between the sun and you, the observer, through an intervening curtain of rain. In other words, you must face the rain with the sun at your back; a rainbow cannot be formed if the sky is completely obscured by cloud.

**The well-known weather proverb, "A rainbow in the morning is the shepherd's warning; a rainbow at night is the shepherd's delight," is reliable about 65 percent of the time.**

If one appears, it will be at a 42° angle up from your shadow.

This position—back to the sun while facing a shower—is the basis for the old weather adage that says a morning rainbow warns of foul weather, and an afternoon or evening rainbow promises clear skies. This bit of weather lore relies on the usual west-to-east movement of weather systems in the mid-latitudes. Thus, if a rainbow is seen in the morning when the sun is to the east of where you stand, then the shower responsible for the rainbow must be to the west, and is likely moving towards you. On the other hand, if a rainbow is in the east in the afternoon or evening, then the rain has passed by and will continue to recede eastward, giving way to clearing skies from the west.

Raindrops act like miniature prisms and mirrors. They split light into a spectrum of colors from violet to red, and then reflect those colors. Whether you see bright bands of color depends on the angle at which the white sunlight strikes the surface of a raindrop. Part of the light is reflected and part of it enters the drop, where it is twice refracted, or bent, and once reflected.

**Rainbows occasionally occur on clear moonlit nights in the same way as daytime rainbows. However, because the light from the moon is dimmer than the sun's, lunar rainbows, or moonbows, are noticeably fainter and more difficult to see.**

The first refraction takes place when the light enters the drop and disperses into colors. On entry, the speed of the ray is slowed; in fact, each color of light is slowed at a different rate and bent at a slightly different angle. Violet light travels most slowly, and thus is curved the most. Red light travels the fastest and is bent the least. When it reaches the opposite side of the drop, most of the light passes through. The remainder is reflected off the inside back of the drop and is refracted again on leaving from the same side it entered. The angle between the entering sunlight and the exiting rainbow rays varies from 42° for red light to about 40° for violet light.

The geometry involved in the formation of rainbows means that those viewed from the ground, when the sun is just above the horizon, will appear as semicircles. The higher the sun is in the sky, the smaller the rainbow. The rainbow disappears altogether for someone on the ground when the sun is at an altitude of 42° above the horizon. A full-circle rainbow is rarely seen, although it may be visible from an airplane when the sun is high in the sky and reflecting off a curtain of rain. Well-defined, bright rainbows are associated with large raindrops. Coloration and definition are generally poor with small drops; for example, fog droplets produce pale, white bows.

**Rainbows always appear in the opposite side of the sky from the sun.**

Each raindrop disperses the full spectrum of colors, but you will only see a single color from each drop, depending on the precise angle of the sunlight and your position. For instance, if orange light from a particular raindrop reaches your eyes, the red from that drop will fall toward the ground in front of you while the green light will pass above your head. You will see green light from raindrops at a lower altitude and blue from even lower drops.

**An icebow is produced the same way as a rainbow, but through ice crystals, not raindrops. Because there are not as many crystals as drops, and the colors are not spread out as much, icebows are generally white. Ordinary snowflakes cannot produce rainbows, so true snowbows are not possible. What we sometimes mistake for snowbows are sundogs.**

It takes millions of falling raindrops—each refracting and reflecting light back to our eyes at slightly different angles—to produce the continuous colored band of a rainbow. As each drop falls, another takes its place in your line of vision, until the number of raindrops begins to diminish and the rainbow fades. The foot of the rainbow is its brightest spot; the seven colors—violet, indigo, blue, green, yellow, orange, and red—bend at that point into a yellowish "pot of gold."

A facet of rainbows that adds to their universal appeal is that each person sees his or her own personal rainbow. While two people may admire the beauty of a rainbow together, what they see is not exactly the same, since each person views sunlight dispersed from a different set of raindrops. Everywhere you move, light from a different set of drops enters your eyes. Chasing rainbows, by the way, does not bring you any closer to them. A rainbow is as near or as far as the raindrops reflecting the sunlight.

**In the tropics, where storms often travel from east to west, afternoon rainbows are often seen before a storm arrives.**

When a bright rainbow is visible, a second, larger rainbow sometimes can be seen parallel to the first. The secondary bow occurs when sunlight enters the raindrops at a specific angle that allows light to make two reflections instead of one at the back of the raindrop. The colors are still refracted at slightly different angles, but those of the secondary bow emerge from the drop at an angle of approximately 51°, not 42° as for the primary rainbow. This is why the secondary bow is larger and positioned above the original bow. It is also fainter because each reflection weakens the intensity of the light leaving the drop. Colors in the outer bow are in the reverse order to those in the primary rainbow, with red on the inside and violet on the outside. Three or more rainbows can occur, but they are usually too faint to be seen. With each additional reflection, there is a reversal of the previous order of colors.

Honolulu, Hawaii, is famous for its brilliant rainbows. At Niagara Falls, visitors are treated to color arcs in the misty spray when the sun is shining. But you don't have to travel far to see a rainbow. Miniature colored arches can be seen in the mist of water sprinklers, fountains, and waterfalls, in the splash of a boat's bow wave, or in the spray thrown up on a wet road. You can even see rainbows inside the drops of heavy dew lying on a lawn or hanging from a spider's web.

# It's Not the Heat, It's the Humidity

Say the word *humidity* and most people think of steamy summer days, long sleepless nights, and perspiring bodies—almost enough to make you long for frosty January nights. These are reactions to the presence of water vapor, that invisible moisture suspended in the Earth's atmosphere. We tend to become aware of water vapor when humidity is high because the air seems sticky, clothes feel damp, and our skin is moist, or when it's low because our lips chap and flyaway hair won't behave. More than most elements in the weather, humidity affects our comfort whether we are indoors or out—nearly always increasing our discomfort as it moves above or below average levels.

**Water moves between the atmosphere and the surface very quickly. On average, a water molecule that enters the air will remain there for about 10 days. By the end of that time, it will have returned to the surface as rain or snow.**

The effect of humidity on humans is almost all bad. High humidity may help to cause migraines, ulcers, clotting, and rheumatoid arthritis, not to mention cramps, irritability, exhaustion, and muddled thinking. Sticky weather affects a person's appearance too: the complexion is flushed, the skin becomes oily, hair is limp, and to make matters worse, humidity usually makes us look and feel bloated.

Excessive humidity can be downright destructive, corroding metals and causing many materials to decompose. Bacteria sprout in high temperatures with high humidities, causing mildew and mold to develop. Potatoes develop blight and wheat rusts, while paper, leather, and manufactured wood products warp and swell. As workers' attention spans shorten, on-the-job accidents and errors increase dramatically. Got a squeaky chair, a guitar that won't stay in tune, or a drawer that sticks? It's not the heat, it's the humidity!

High humidity also influences respiration and perspiration in animals. Customs agents in Saudi Arabia found that when the weather gets too damp, narcotic-detector dogs become less effective. In heat and high humidity, farmers claim hogs grow restless, lose their appetites, and gain less weight, and cows milk poorly and conceive fewer offspring.

**During the 1996 Summer Olympics in Atlanta, officials prepared for high heat and humidity by commissioning a company to set up misting fans in canopies. Tiny nozzles around the fans sprayed droplets of water that evaporated, thus cooling the air by 20–30°F. Both athletes and horses chilled out this way. Officials had good reason to be concerned: Atlanta is typically very hot and humid in the summer. The average relative humidity on a July morning in Atlanta is 88 percent, and the normal daily maximum in July is 88°F. Under these conditions, the heat index would be about 115°F, uncomfortable for anyone, especially an athlete in competition.**

Humidity is closely linked to human discomfort because of its effect on perspiration. In hot weather, the body regulates its core temperature by sweating. Sweating itself doesn't cool, because the temperature of perspiration is the same as that of the body. As the sweat evaporates, however, it carries heat from the body, cooling the skin. This evaporative process works best when the air is dry. As humidity rises, the moisture-laden atmosphere cannot take as much water away from the skin and absorbs it too slowly to cool efficiently. In extreme conditions, when the body loses large amounts of fluid through perspiration, heat exhaustion and heat stroke may occur. Symptoms include profuse sweating, weakness, nausea, muscle cramps, headache, faintness, and feverishness.

Although invisible, water vapor is always present in the atmosphere, varying in concentration from almost zero to a maximum of about four percent by volume. Water vapor is a vital element in meteorology. It is the only gas that can change into a liquid or solid under ordinary atmospheric conditions, and it is this changeability that results in many features of the weather—clouds, dew, rain, snow, fog, and frost.

Warm air can hold more water vapor than cold air. If a specific volume of air at 32°F can hold a quarter of a quart of water, it can hold a half a quart at 50°F and two quarts at 91°F. Air is said to be "saturated" when it is holding all the water vapor it can contain for the prevailing temperature and pressure.

Meteorologists have coined several expressions to describe the air's moisture content. The most familiar, and perhaps the most misunderstood term, is *relative humidity*. It means the percentage of water vapor that air actually holds compared with what it could hold at a specific temperature and pressure. It is frequently misunderstood to mean the air's moisture content, or absolute humidity. Relative humidity varies from 100 percent in clouds and fog to 10 percent or less over deserts during the day. Suppose air at a specific temperature can hold about a quarter of an ounce of water vapor per cubic yard at its capacity. If, however, it contains only an eighth of an ounce, then it is only half saturated, and its relative humidity is 50 percent.

**A few centuries ago, Roger Bacon discovered that a pound of wool hung suspended for a single night over, but not touching, the surface of the water in a well will have absorbed so much of the dampness by morning that its weight increases by slightly more than a pound.**
**A rope knot tied in dry weather will become harder to untie in damp weather. Old-time sailors anticipated a storm when knots began to tighten.**

By itself, relative humidity tells us nothing about how much water is in the air. A high relative humidity does not necessarily mean high humidity. Relative humidity follows air temperature in an inverse way—a decrease in temperature results in an increase in relative humidity, and an increase in temperature causes a decrease in relative humidity. For example, a relative humidity of 70 percent feels more humid when the temperature is 50°F than when the temperature is 70°F. On average, the relative humidity is greatest at dawn, the coolest part of the day, and lowest in midafternoon, when the temperature reaches its maximum. It tends to be highest in winter and lowest in summer.

A much better term to describe atmospheric moisture is *dew point*. That is the temperature at which an air mass becomes saturated, usually by cooling. While the air temperature remains above the dew point, the

**National Weather Service's Chart of Heat Index Risks**

| Heat Index (°F) | Possible effects |
|---|---|
| 80° to 90° | Fatigue possible with prolonged exposure and/or physical activity |
| 90° to 105° | Sunstroke, heat cramps, and heat exhaustion possible with prolonged exposure and/or physical activity |
| 105° to 130° | Sunstroke, heat cramps, and heat exhaustion likely, and heatstroke possible with prolonged exposure and/or physical activity |
| 130° and higher | Heatstroke or sunstroke very likely with prolonged exposure |

Source: National Weather Service

air is unsaturated, that is to say, it is capable of holding additional water vapor. If the air temperature near the ground is 73°F and the dew-point temperature is 55°F, the air must cool 18°F to reach saturation. Once it reaches the saturation (its dew-point temperature) it releases its excess moisture by forming clouds, rain, or fog, or by condensing as dew or frost on cold surfaces, such as grass or vehicles. The dew point does not change with the air temperature. Over the course of a summer day, while relative humidity is rising and falling, the dew point will remain fairly constant as long as the moisture content remains unchanged.

**Performance testing shows that industrial accidents and stenographic errors increased by a third when humidity reached oppressive levels.**

The higher the dew-point temperature, the more moisture in the air. With a dew point of 68°F, most people feel uncomfortably sticky, even though the relative humidity may only be 50 percent and the air temperature 90°F.

Over the years, climatologists have proposed several terms for the various combinations of temperature and humidity that are used to describe what hot, humid weather feels like to the

**Heat Index Chart**

| Relative humidity (%) | Air temperature (°F) 70 | 75 | 80 | 85 | 90 | 95 | 100 | 105 |
|---|---|---|---|---|---|---|---|---|
| 0 | 64 | 69 | 73 | 78 | 83 | 87 | 91 | 95 |
| 10 | 65 | 70 | 75 | 80 | 85 | 90 | 95 | 100 |
| 20 | 66 | 72 | 77 | 82 | 87 | 93 | 99 | 105 |
| 30 | 67 | 73 | 78 | 84 | 90 | 96 | 104 | 113 |
| 40 | 68 | 74 | 79 | 86 | 93 | 101 | 110 | 123 |
| 50 | 69 | 75 | 81 | 88 | 96 | 107 | 120 | 135 |
| 60 | 70 | 76 | 82 | 90 | 100 | 114 | 132 | 149 |
| 70 | 70 | 77 | 85 | 93 | 106 | 124 | 144 | 163 |
| 80 | 71 | 78 | 86 | 97 | 113 | 136 | 157 | 180 |
| 90 | 71 | 79 | 88 | 102 | 122 | 150 | 170 | 199 |

Source: National Weather Service

average person. The *Heat Index*, also called apparent temperature, is widely used in the US to measure the discomfort that humidity adds to temperature. It originated in 1979, when R. I. Steadman developed an index with many factors. The Heat Index is a simplified version of Steadman's version.

The Heat Index is calculated using relative humidity and air temperature, or dew-point and air temperature. An air temperature of 85°F with a relative humidity of 80 percent results in a heat index of 97°F—that is, it feels like 97°F. At this temperature, sunstroke, heat cramps, and exhaustion are possible. If the temperature is 85°F and relative humidity is 100 percent, the heat index is 108°F. At this apparent temperature, sunstroke, heat cramps, and heat exhaustion are quite likely to occur. Some individuals might experience heatstroke at this apparent temperature. Of course, comfort is subjective and largely dependent on the age and health of the individual. Weather conditions causing prickly heat in an infant may result in heat cramps in a teenager, heat exhaustion in a middle-ager, and heat stroke in a senior. The Heat Index is also limited as an overall hot-weather comfort index because it does not consider other factors such as pressure,

**Zookeepers report very few fatalities during heat waves, primarily because animals know what to do. Nothing!**

wind speed, precipitation, sunshine, or pollen. For example, in direct sunshine, the Heat Index can rise as much as 15°F.

Prolonged high humidities are common in many parts of the US, particularly the South and Southeast. In many cities in the US, the deadliest weather is high heat and humidity. According to the National Weather Service, about 175 Americans die from the heat in an average year. From 1936 to 1975, some 20,000 people in the US died from the heat. During the heat wave of 1980, over 1,250 died, and more recently, in 1995, 583 people died in a heat wave in Chicago. Generally, the Heat Index decreases the farther north you go, but that doesn't mean that Northerners don't have to be careful. From July 13–15, 1995, 57 people died in Wisconsin from the heat. The temperatures there ranged between 100°F and 109°F, and the dew point was around 80°F, resulting in a Heat Index as high as 130°F. People usually acclimatize to heat and humidity, which is why sometimes those living in areas that are not typically hot and humid—like Wisconsin, where a typical July day is about 70°F—are especially prone. Among the most vulnerable, regardless of region, are the elderly. Those living in cities are at especially high risk, since cities can become "heat islands." Cities are also dangerous because stagnant air in hot and humid conditions leads to higher pollution levels.

**Moisture can cause musical instruments to go out of tune. That's why they say that if a fiddle won't stay in tune, there's going to be a storm. However, a UCLA study suggests that the best weather for a concert is cool and humid: sounds last longer in cool, humid air than in hot, dry air.**

Having a bad-hair day? It may be the humidity. The sensitivity of hair to humidity has been scientifically recognized since 1783, when a Swiss physicist, Horace de Soussure, found that hair stretches by about 2.5 percent when air goes from complete dryness to saturation. He developed the hair hygrometer for measuring atmospheric moisture. Strands of hair or a single hair are firmly supported by a mechanical linkage system that is also attached to a pen or pointer arm. As hair in the hygrometer lengthens and shortens, it moves the arm across a chart.

Over the years, hygrometers have used many organic materials such as skin, sheep gut, and hemp rope, but animal and human hair were found to work best. Blond or red hair were preferred, because they are even more responsive than brunette or black hair. During World War II, the US War Department advertised for "blonds with hair of soft texture and the highest quality, not less than 12 inches long—all in the interest of national defense." Almost 200 miles of blond strands were needed in humidity instruments being sent aloft in upper-air balloons. Until then, hair used in hygrometers had been imported from the Balkans.

Because hair responds very slowly at low temperatures, and not at all below −0°F, the hair hygrometer is not widely used anymore. Today, most meteorologists use the psychrometer. It consists of two thermometers, one an ordinary dry bulb and the other a dry bulb wrapped in a moistened gauze sleeve. Water evaporating from around the wet bulb makes it cooler than the dry bulb. The difference in the two temperatures gives a measure of the relative humidity and dew point temperature, which can be obtained by looking them up in psychrometric tables or charts.

# North American Weather Moments

# Let It Snow!

## The Blizzard of 1888

New York shopkeepers advertised "Spring Opening Day" for Monday, March 12, 1888. After all, Friday, March 8, had been the warmest day of the year, and New Yorkers were enjoying balmy weekend temperatures in the fifties. Walt Whitman, the *New York Herald's* staff poet, wrote a poem which concluded with: "The spring's first dandelion shows its true face."

But on Monday morning when New Yorkers awoke, it was snowing.

Heavy, white flakes blew through New York all day long. Fierce winds lashed the city in gusts of up to 80 mph. Snowdrifts piled up as deep as 20 feet. By that afternoon, the city that considered itself the most sophisticated in the world looked as though someone had smothered it under a monstrous, fluffy white quilt.

The power and telephone wires were down. Thousands of people were stuck in unheated elevated trains and horse-drawn carriages. The ferries stopped running. Wall Street stockbrokers went home—it was the first time the New York Stock Exchange had been closed because of the weather since its opening in 1790. Most theaters canceled their performances. New Yorkers found themselves as isolated as a small ship in a stormy ocean—though being at sea at the time was worse than being blanketed under the snow; some 200 ships were wrecked off the coast that day, like toys in a bathtub.

**The show must go on... During the infamous Blizzard of '88, New York came to a standstill. But Daly's Theatre didn't disappoint theater-goers who battled 20 inches of freshly fallen snow and howling winds to get to the Shakespeare play that evening. The play? *A Midsummer Night's Dream.***

Frozen bodies and dead horses were pulled from the snowdrifts. In the city, 200 people perished in the storm. Another 200 died in other stricken areas of the Northeast. Washington and Philadelphia were also crippled by the great blizzard.

The Blizzard of '88 humbled New Yorkers. It led them to build an underground subway and underground telephone wires. But most of all, it made them remember how vulnerable humans are to nature's wrath, and how we have to adapt to its unexpected fury.

That was over a century ago, and though technology has brought us many gifts to cope with winter storms and cold weather—electric snow blowers, studded tires, polar fleece clothing, cellular phones, battery-operated flashlights, indoor shopping centers—snowstorms are still a humbling experience for the millions of North Americans who get stuck in them each year.

## The Blizzards of 1993 and 1996

According to the National Weather Service, a snowstorm becomes a blizzard when the wind reaches 35 mph, the temperature plunges below 20°F, and visibility is limited due to blowing snow.

Most snowstorm fatalities are indirect: people die in car accidents, of heart attacks while shoveling snow, and of exposure to the cold.

One of the more devastating snowstorms in recent memory whipped the whole eastern seaboard from March 12–15, 1993. It was called the "Storm of the Century" and its snow and high winds buried the East. Snowfalls of up to two feet were common, and the storm shut down airports from Maine to Georgia. Half of the states in the country were affected. Three million utility customers lost power. By the time the storm was over, some 270 people had died, another 48 lives were lost at sea, and damages were estimated at between $3 billion and $6 billion.

All along the coast, powerful waves hammered the shore, and 18 waterfront Long Island homes fell into the sea. In Florida, a 12-foot storm surge killed seven people, and a fierce outbreak of 50 tornadoes took 18 lives.

Record amounts of snow fell in the South. Seventeen inches of snow fell near Birmingham, Alabama, and the wind blew it into drifts up to six feet high. Snow fell on "every square inch" of Alabama. In Georgia, over 1.3 million unlucky chickens died. Some parts of Atlanta had nine inches of snow, and the city was issued one of the first blizzard warnings in its history. Up to half a foot of snow blanketed Florida's panhandle. Chattanooga, Tennessee, which normally gets a

## Snowiest Places

- On average, Valdez, Alaska, gets more snow each year than any other US town or city: 325.8 inches.
- An annual average of 253.9 inches of snow falls on Mount Washington, New Hampshire.
- Marquette, Michigan (population: 22,000), receives 130.6 inches of snow in a typical year.
- The snowiest city with a population over 100,000 in the US is Syracuse, New York (population: 165,000), which gets 114.7 inches each year.
- The snowiest state capital is Juneau, Alaska, with an annual average of 100.7 inches of snow.
- Among the snowiest capitals in the lower 48 states are Albany, New York, with an average of 63.8 inches a year; Concord, New Hempshire, with 63.5 inches; and Denver, Colorado, which gets 60.3 inches in a typical year.

total of 4.3 inches of snow all year, was smothered under 20 inches. Four feet of snow fell in the Great Smoky Mountains of Tennessee, and 200 hikers had to be rescued by helicopter. It was one of the largest peacetime rescues in the US.

Syracuse, New York, received 43 inches of snow. In New York City, howling winds shattered windows from skyscrapers and sent glass daggers plunging to the streets below. Wind lashed Mount Washington, New Hampshire, at gusts of 144 mph. In Burlington, Vermont, the mercury dipped to a frosty −12°F. Boston had to cancel its St. Patrick's Day Parade for the first time in its history.

The "Storm of the Century" ranked as one of the strongest storms on record, with record-low barometer readings observed in many states. Low pressure was blamed for an unusually high outbreak of childbirths.

And the amount of snow that fell was huge. The Office of Hydrology of the National Weather Service estimated that the amount of snow that fell, if melted, was about 40 times the volume of water that flows out of the Mississippi River at New Orleans each day.

### Southerly Snows

- Snowflakes fluttered from the sky at Homestead, Florida—about 30 miles south of Miami—on January 19, 1977.
- Over a foot of snow fell on San Antonio, Texas, on January 12, 1985. Normally the city gets a total of less than an inch per year.
- The Texas panhandle received as much as 33 inches of snow during a storm that started on Groundhog Day 1956.
- Tuscon, Arizona, was blanketed with 6.4 inches of snow in November 1958. It snowed again in Tucson on Christmas Eve and Christmas Day 1987—it was the first white Christmas there in 47 years.
- El Paso, Texas, was walloped with 16.8 inches of snow in 24 hours on December 13, 1987. By the time it stopped snowing, the city was smothered under 22.4 inches.
- In 1989 a cold snap brought a white Christmas to northeastern Florida, Georgia, and the Carolinas—the first one on record.
- New Orleans was covered with 4.5 inches of snow on New Year's Eve 1963. Elsewhere in the South, Meridian, Mississippi, was clobbered by 15 inches of snow—heavy for a city that averages just over an inch of snow in a whole year.
- It snowed in Hawaii in 1995. Up to six inches of snow fell on the summits of Mauna Loa and Mauna Kea on April 24.

But if the 1993 blizzard was the storm of the century, what were people to call the one that hit in 1996?

Snow started falling in the Eastern states in the afternoon on January 6, 1996. By the time it stopped early in the morning on January 8, the East was, yet again, buried in up to two feet of snow. The Blizzard of '96 couldn't really be dubbed the "Storm of the Century"—just three years after the other one—but it was just as extraordinary, and it crippled the East just as thoroughly. Snow fell in record quantities in parts of New Jersey, Maryland, Pennsylvania, the District of Columbia, Delaware, Virginia, West Virginia, Connecticut, and Ohio. In fact, most records

## Famous Snows

- George Washington noted three feet of snow at Mount Vernon on January 27, 1772. Meanwhile, Thomas Jefferson reported the same amount at Monticello.
- Ninety-six movie-goers died in Washington, DC's Knickerbocker Theater when heavy snows caved in the theater's roof on January 22, 1922.
- A snowstorm brought about the first case of cannibalism tried in the US. In January 1874, Alfred Packer led a team of prospectors into the San Juan Mountains of Colorado. Initially there were 21 in the party, but at a camp near Montrose, the group was advised not to try to cross the mountains because of snow. Packer and five others went anyway, and they got lost and snowbound sometime in February or March. In April, Packer showed up at the Los Pinos Indian Agency, alone. The others were never seen again, though human flesh was found in August on the trail. Alfred Packer (sometimes spelled Alferd) was eventually tried for cannibalism and sentenced to be hanged, but later his sentence was reduced to prison time. One serious accusation raised against him at the time was that he had eaten most of the Democrats in the county. Some say that after the trial, Packer became a vegetarian. His legacy? He's a Colorado folk hero, and a cafeteria at the Boulder University Memorial Center is named the Alferd Packer Memorial Grill.

that had been set in the "Storm of the Century" in 1993 were easily broken in January 1996.

The human and economic toll was severe: 187 people dead and some $3 billion in damages. One economist estimated that lost production and sales totaled around $17 billion. All along the coast, airports were closed and hotels were vacant. Thousands of flights were canceled and delayed, and luggage was taking off without passengers.

"Assume bags will be lost," one newspaper advised, "and pack a change of clothes and other essentials in a carry-on." One thousand

## Animals in the Snow

- On March 8, 1717, 1,200 sheep were found trapped in a snowdrift on an island off the coast of New York. They had been there for four weeks, but 100 of them were still alive.
- Four out of every five cattle in Kansas died in a raging blizzard on January 13, 1886.
- A three-day blizzard that started on January 28, 1887, wiped out millions of free-range cattle in Montana. Cowboy "Teddy Blue" reported that by the time the "Big Die-Up" was over, more than 60 percent of Montana's cattle were dead. The snowy, cold winter hastened the end of the Old West's open range system, as people moved toward homesteading and ranching.
- One million Thanksgiving turkeys froze to death in a South Dakota blizzard on November 18, 1943.
- Over a foot of snow fell in Arkansas on January 6, 1988, killing 3.5 million chickens. Meanwhile, another 1,750,000 chickens died across the border in Texas. The following day two million more chickens died in Alabama.

passengers had to camp out in Miami airport, waiting for flights to the Northeast. The Red Cross reported blood shortages. Bread and milk were in short supply: one grocery chain reported that customers instantly bought all the loaves of bread from the shelves as soon as they were stocked.

People stayed at home: cyberspace traffic was way up. The use of America Online was 60 percent above normal.

The storm hit Washington, DC, the day before federal employees were supposed to return to work, after being furloughed for three weeks because Congress had trouble passing the budget. Almost 12 inches of snow fell on Sunday, January 7. Workers were told to stay home from Monday through Wednesday. They returned to a flurry of work on Thursday, and most departments reported huge backlogs. At the Passport Agency, frustrated travelers stood in line waiting for their travel documents; the agency reported a backlog of 200,000 applications. And just

## California Snows

- On February 5, 1887, it snowed almost four inches in San Francisco.
- On January 10, 1949, it snowed in San Diego. On the same day, Burbank was hidden by what for Southern California amounted to a major snowstorm: 4.7 inches.
- Sacramento was covered with two inches of snow on February 5, 1976. Its record was 3.5 inches on January 4, 1888.
- On December 16, 1987, it snowed on Malibu Beach. In nearby Anaheim, Disneyland was shut down due to the weather.
- The highest monthly snowfall in the US was recorded in Tamarack, California, in 1911, when 369 inches fell throughout January. This contributed to the highest snow depth ever recorded in the US, with 451 inches covering the ground in Tamarack by March of that year.
- A storm on October 28, 1846, left five feet of snow on a pass high in the Sierra Nevada between Reno and Sacramento. It stranded a caravan of settlers led by George Donner. They were stuck for the winter, and of the 79 people who had set out on the expedition, only 45 survived. There were stories of cannibalism. The pass is now called Donner Pass. Another blizzard in February 1936 at Donner Pass left 750 motorists stranded. Seven people died.
- At Tahoe, California, 149 inches of snow fell in a single storm in 1952. Seven years later, 189 inches fell in a single storm at Mount Shasta Ski Bowl on the northern edge of California.

when workers thought they were back on the job, another storm of snow, ice, and drizzle hit Washington on Friday and shut down offices again.

Over two feet of snow covered Dulles International Airport. In the capital region, which received more snow in those two days than it usually gets in a whole year, it was better to stay at home than to try to drive through the snowy streets or take a crowded train. A subway train in the suburbs slid and crashed into another train; the driver died. Most highways in the region were shut down.

A rare snowfall covers Santa Fe, New Mexico. *Corbis*

- Blue Canyon, California, northeast of Sacramento, gets an average of 241 inches of snow—one of the snowiest places in the country.
- Five feet of snow fell in 24 hours at Giant Forest, in Sequoia National Park, in 1933.
- Southern Californians basked in 80° heat while the Blizzard of '96 blasted the East Coast, but just seven years earlier, in 1989, Los Angeles had had to cope with its own snowstorm. Cars clogged the freeways in Los Angeles, and there was even snow in Malibu.

In Virginia, snow removal costs were about $50 million. Twenty inches covered Lynchburg in just 24 hours. In Maryland, it cost more than $70 million to remove snow and fix damage to roads. One hundred cows died and 100 more were hurt when heavy snow collapsed the roof of their Frederick County barn.

New York was buried by over 20 inches of snow. Flights were delayed and canceled. Passengers on United Flight 801, which was supposed to fly from Kennedy Airport to Tokyo, were stranded in the plane

## Summer Snows

- On June 11, 1842, a snowstorm hit New England. Up to a foot of snow fell in parts of Vermont. As much as a foot and a half of snow fell on Dickenson Park, Wyoming, on June 24, 1989. July 1883 was a snowy month for Pike's Peak, Colorado. A total of 23 inches fell throughout the month. Up to a foot and a half of snow fell near Mirror Lake, Utah, on July 23, 1993. The northeastern part of Yellowstone Park was covered with up to four inches of snow on August 16, 1987. In the summer of 1992 it snowed more than eight inches in Great Falls, Montana, on August 22 and 23.

for 7.5 hours on the runway; one passenger later reported that the captain threatened to arrest people because they were getting so riotous. The governor declared a state disaster and requested help from the Federal Emergency Management Agency. The public school system was shut down for the first time in almost 20 years.

Boston received 18 inches of snow, bringing the snow depth there to a record 32 inches. A Boston Bruins game was rescheduled. One Logan Airport hotel planned a blizzard party.

Philadelphia struggled with about 30 inches, 27.6 of which fell in just 24 hours. The storm brought the city to a standstill. The US Post Office failed to deliver on Monday, despite its familiar slogan: "Neither snow, nor rain, nor heat, nor gloom of night, stays these couriers from the swift completion of their appointed rounds." (Within a couple of days, though, the hardy workforce was delivering 90 percent of the city's mail.) Schools and courts were closed. Snow accumulation on roofs was so heavy that many collapsed. A greenhouse crumbled under the snow's weight and killed a man inside.

Twenty-nine people died of heart attacks in Pennsylvania on January 8 as a result of shoveling driveways. And snow shoveling was dangerous in other ways: a 70-year-old man in an Allentown suburb accused his neighbor of shoveling snow onto his car; the angry neighbor pushed him down and he died. Overall, 99 Pennsylvanians died.

Eighty of the deaths were due to the blizzard; the rest died in the subsequent flooding.

Two families were left without their home when snow in the streets prevented firefighters from getting to a burning house. Huge icebergs formed in the Schuylkill River when the city dumped 2,000 tons of snow in the river. Two hundred snowplows tried in vain to clear the airport runways. The Harley-Davidson plant in York, Pennsylvania, shut down.

The blizzard wasn't bad news for everyone. A ski resort in Pocahontas County, West Virginia, basked under four feet of accumulated snow.

When a blizzard hits, it seems that everything turns white. Even the last resort of the desperate—crime—sometimes gets distinctly snowy. In New York during the Blizzard of '96, it was reported that two armed gunmen held up and robbed a Bronx building superintendent. But the thieves weren't after his money. They took his snow blower.

## North America's Coldest Day

On February 3, 1947, at 7:20 a.m. local time, the weather observer Gordon Toole hurried the 100 feet from the warm log barracks at Snag Airport, in Canada's Yukon Territory, to the weather-instrument compound next to the runway. For eight straight days the temperature had been below −58°F, but on this morning it felt colder. Toole could plainly hear the dogs barking in the village three miles to the north, and his exhaled breath made a tinkling sound as it fell to the ground in a white powder. His six husky dogs were really feeling the cold. They were asleep on top of their kennels, curled up with their noses tucked right up under their tails to garner every calorie of heat.

By the time he arrived at the white-louvered shelter housing the thermometers, he could feel the cold seeping through his parka. He unlatched the door of the instrument shelter and shone the flashlight inside, but was careful not to lean forward and breathe on the thermometers. He saw something that he had never seen before: the tiny

**"Besides the two horses . . . [and] sled dogs, there were ravens, rabbits, and ptarmigan trying to survive the cold. Many mice also sought refuge in our warm buildings. The janitor had a large tomcat, so the poor unfortunate mice didn't fare so well! Needless to say, the cat was very happy and well fed! The horses, with their two- to three-inch buildup of snow and ice on their hoofs, didn't seem to suffer from the cold, and our cook found it in his heart to feed them a variety of 'snacks,' i.e., wieners, peelings, crusts of bread, etc. . . . Our most memorable day was the day we watched the DC3 land with new supplies, vegetables, meat, fruit, and, of course, a few cases of beer and liquor. We certainly played some high-stakes poker that night! We had lots of interviews and pictures taken that day. The next day we had some pretty severe hangovers."**
**–Wilf Blezard, Weather Observer at Snag, Yukon**

sliding scale inside the glass thermometer column had fallen into the bulb at the end, well below the −80°F point—the last mark on the thermometer.

Toole rushed back to the barracks where he coaxed his colleague, Wilf Blezard, to return to the instrument compound. Using a set of dividers to measure the tiny bit of alcohol left in the column, Toole estimated the temperature to be about −83°F. As he dutifully scratched a mark on the outside of the thermometer sheath adjacent to the end of the alcohol, he thought about what the Canadian Weather Service head office had advised three days earlier. If the alcohol level ever fell below −80°F, they should mark a corresponding point on the thermometer sheath with a pen. Typical advice from head office, Toole thought: ink does not flow at that temperature. Instead, he made the historic mark using a fine, sharp file.

To complete the job, the observers noted the weather was a repeat of the past two months—clear, dry, and calm. Snow on the ground amounted to 15 inches, but was evaporating at over an inch a day. The visibility at eye level was 20 miles; however, on February 3, ground visibility was greatly reduced. At about arm's length, an eerie, dull gray shroud of patchy ice fog hung above the dogs and heated buildings.

Back inside the Snag weather office,

the radio operator transmitted the weather observation to Whitehorse and Toronto. Within the hour, the director of the Canadian Weather Service congratulated Snag on becoming North America's "cold pole." He also asked Toole to send the thermometer back to be recalibrated. The two observers shared the news with the rest of the camp before packing the thermometer for air shipment to Toronto to have the readings confirmed.

But almost a week passed before it was warm enough for an airplane to land at Snag. Once in Toronto, the thermometer was put through several laboratory tests before technicians concluded that it had been reading about 1.6°F in error. Three months later, the weather service accepted a value of −81.4°F as the corrected temperature—still the lowest official temperature ever recorded in North America. It is a record that still stands today—more than 50 years later.

**The coldest Inauguration Day ceremony in the US was the one marking the beginning of Ronald Reagan's second term, on January 21, 1985. The wind chill factor made Washington, DC, feel like −30°F.**

At Snag that day, the 16 airport employees did not need confirmation in Toronto. They could feel how cold it was. Still, they were excited by the news. Blezard, now retired, recalls, "We had to put a little lock on the door to the instrument screen because everybody was rushing out and looking at the thermometers. Even the slightest bit of body heat would cause the alcohol to jump." Fifty years is a long time ago, so perhaps it is not surprising that Toole's memories of the day are different. "Staff interest," he said, "was pretty limited. There was no euphoria, prolonged celebrating, or serious discussion on how to commemorate the moment."

Perhaps no one understood the historic significance, or maybe it was just that the cold showed no sign of abating. But that was to change. By 2 p.m., the day's high had reached a relatively balmy −54°F. Before the day was over, media from around the world had besieged the "frozen chosen" for exclusive interviews on the historic cold. Writers from the *Milwaukee Journal* and *Vancouver Sun* phoned for front-page stories to learn what −80°F felt like. One newspaper carried the following

Observing air temperature at an Arctic weather station. *Environment Canada*

headline: "Snag snug as mercury [sic] sags to a record −82.6°F." The thermometers did not use mercury because it freezes at −39°F.

They were alcohol thermometers and the newspaper's reporter, although not the headline writer, knew it. Later in the story: "The only reason the men didn't celebrate was that all the alcohol at the station was in the thermometer and that was nearly frozen."

Telegrams of congratulations arrived from many countries. But some experts expressed skepticism. For British meteorologists, who at that time measured coldness in degrees of frost, upward of 115 degrees of frost was just too much to comprehend. (Degrees of frost refer to the number of degrees the temperature falls below the freezing point of 32°F.)

On February 8, a plane arrived at Snag with American military and media who wanted to learn what it was like living and working in such cold conditions. The men at Snag, however, were more interested in the visitors' cargo of meat, beer, and rye than in becoming celebrities to strangers. Hearing from family and friends was different. "We were celebrities to them," remembered Toole. "And finally, they could locate Snag on the map of North America."

Snag was named during the Klondike gold rush. Because boatmen could not read the silty waters of nearby stretches of the White River and its tributaries, boats had to be poled going upstream. On occasion, they would be "snagged" and punctured by sharp, pointed tree trunks submerged below the milky waters, hence the name.

A long-time Yukon resident, Jean Gordon, of Mayo, about 200 miles northeast of Snag, knows many hunters who have lost all their supplies, guns, and meat when a snag speared their boat. She explains that snags come from trees that have washed down from riverbanks during high water. They move downstream with the heavy butt and root end sinking and the treetop facing downstream. As the water recedes, the butt will become anchored to the stream bed and the rest of the tree will oscillate in the water. As the limbs become denuded, a long snag pointed downstream, which can move up and down in the current, creates a very dangerous hazard, especially for boats traveling upstream with the snag below the surface. Experienced river men can read the water even when it is muddy or full of silt, but they hate waters where there are snags.

**Based on normal daily mean temperatures in January, the coldest state capital in the US is Bismarck, North Dakota, with an average January temperature of 9.2°F.**

The Snag weather station operated from 1943 to 1966. It was located at Snag Airport, east of the Alaska-Yukon boundary, and 15 miles north of the Alaska Highway at Mile 1,178. The airport was at coordinates 62° 23'N and 140° 23'N, with an elevation of 2,120 feet. Set in a broad, bowl-shaped, north-south valley of the White River, a tributary of the Yukon River, the now-abandoned airport was surrounded by unglaciated uplands of moderate relief. The vegetation was mostly scrub and poplar trees about 10 to 20 feet tall. The magnificent St. Elias mountains lay 30 miles to the south. The tiny village of Snag, consisting of less than a dozen people, was about 3.5 miles to the north of the airport, near the point where the Snag Creek flowed into the White River.

Of the 16 staff members at Snag Airport, four single men in their early twenties were there to observe the weather. Toole was the officer in charge of the weather station. Meteorological staff earned about $1,320 annually, with an extra $20 monthly isolation allowance, which covered the room rate in the barracks. The daily food charge was $0.50.

The other airport employees were radio operators, employed by the Department of Transport, and airport maintenance and operations personnel, employed by the Royal Canadian Air Force (RCAF), whose main job was to keep the runway open. Incidentally, in winter this meant compacting the snow, not plowing or blowing the strip bare. Snag was part of the Northwest Staging Route—one of several emergency landing strips or observing stations connecting the Yukon and Alaska with central Canada and the United States. They were set up in 1942 and 1943 to provide basic weather services for the RCAF, the United States Army Air Force, and civilian aviation companies providing military transport. Most pilots flying the northwest route had to fly with visual contact with the ground, called visual flight rules (VFR); otherwise, they might get lost. If weather socked in the main airports, the pilots used alternative airports like Snag and Smith River.

**Based on normal annual temperatures, the coldest states are Alaska, North Dakota, Minnesota, Maine, Montana, and Wyoming.**

On February 3, the thermometer at Fort Selkirk, a very small community on the Yukon River, over 300 miles east-northeast of Snag, recorded −85°F, corrected for instrument error. This reading, however, was not considered official because the thermometer was exposed on the outside wall of a building and not housed in the standard instrument shelter. That same day at the station at Mayo, the temperature apparently reached −80°F. "Apparently" because at midnight on February 15, the station burned down, destroying the weather instruments and observation records. Nevertheless, photographic evidence shows Mayo's temperature reading on that day as about −80°F, just marginally above the Snag low. Mayo booster Jean Gordon claims that while Snag can claim a lower temperature extreme, Mayo, with its two schools, hotels, and population of 500 people, can boast being the coldest "decent-sized" community in North America. As a road sign entering Mayo asserts, it's also the town with the largest temperature range: a huge 177 degrees, from a maximum of 97°F to a minimum of −80°F.

How did such cold happen? As in most Arctic cold spells, weather conditions in 1946–47 were favorable for a steep temperature inversion. Inversions, a frequent feature of Arctic winters, are exceptions to the general rule that temperature decreases with increasing altitude. Inversions can be produced by gravitational drainage of cold air or by radiation. In elevated terrain, the heavy, dense air sinks and slides down the mountain slopes, often pushing any warmer air aloft. The ground also grows colder by radiating heat to the cloud-free sky. In doing so, the ground readily cools the air immediately above it, especially when the skies are clear, there is unlimited visibility, and the winds are calm or light. A layer of air closest to the ground may be as much as 20°F to 40°F colder than the air at 3,000 feet.

**How cold can it get in Hawaii? On May 17, 1979, a temperature of 12°F was observed at Mauna Kea Observatory. And Florida? It was -2°F at Tallahassee on February 13, 1899.**

In 1946–47, a strong westerly circulation across North America confined cold Arctic air over Alaska and northwestern Canada for much of the winter. During this time, the cold dome of heavy, dense air over the Yukon intensified. With a continuous supply of cold air from northeastern Siberia, the cold dome over the Yukon grew in extent and severity, creating all the record lows. But a dramatic change was to occur later in February: the westerlies relaxed, the cold air spilled through to eastern North America, resulting in severe cold as far south as Florida, and brought maritime air from the Pacific to the southern Yukon, where the cold broke for a few days. At Snag, the temperature even rose to a more civilized 45°F.

How did −81.4°F feel? Most North Americans never experience temperatures lower than −50°F. Blezard and Toole repeatedly said there was a considerable difference between −50°F and −80°F.

The following anecdotes, pieced together from station correspondence, as well as from recent interviews and correspondence with the two observers, give us a glimpse of what life was like for the frozen 16 at Snag during the winter of 1946–47.

Says Blezard: "An aircraft that flew over Snag that day at 10,000 feet was first heard when over 20 miles away; and later, when overhead, still at 10,000 feet, the engine roar was deafening. It woke everyone who was sleeping at the time, because they thought the airplane was landing at the airport."

Anyone who has ever skated outside or gone for long walks in the dead of winter knows that the colder it gets, the farther sound carries. That is because sound ordinarily spreads obliquely upward over our heads and is therefore not heard very far away. But in very cold, stable air, the inversion bends the sound waves back towards the earth, where they tend to hug the ground. Furthermore, audibility is improved by the absence of turbulence or wind. In the end, conversations usually heard 100 feet away can be heard more than half a mile away if the air is clear.

**Modern technology has given us many ways of coping with the cold, but things weren't so easy last century. When pioneers and homesteaders got caught in severe cold, sometimes that was the end. During a cold snap across the Great Plains on January 12, 1888, over 200 pioneers and thousands of cattle died.**

However, at temperatures below −58°F, ice fog reduces the transmission of sound waves considerably. Here is an excerpt from a letter the officer-in-charge at Mayo, Matthew H. "Harry" Ewing, wrote on March 8, 1947, to Dr. Andrew Thomson, director of the Canadian Weather Service in Toronto. In it, he commented about a friend of his, a trapper and prospector in the hills north of Mayo, who claimed that ice fog definitely kills sound, whereas severe cold without fog is very favorable for conducting sound:

"Once the old-timer was working across a valley three or four miles from another party," Ewing wrote. "The valley was filled with fog at 60°F below but there was no fog on the mountains. In the very cold air, when the men across the valley were chopping wood and making other noises around camp, he could hear them very clearly, much clearer than normally."

At such temperatures, there are even extraordinary sounds: for instance, staff at Snag could not only see but also hear their moist breath solidify to ice outside, in a hissing or faint swishing sound. A piece of thin ice, when broken, sounded exactly like breaking glass. From his home in

Watson Lake, Yukon, Toole recalled: "Ice in the White River, about a mile east of the airport, cracked and boomed loudly, like gunfire. During the bitter cold, you would go days without seeing any wildlife, apart from ravens, rabbits, mice, snowbirds, and ptarmigans. Cold air generated intense radio static much like the crackling during a thunderstorm."

There were other cold-weather experiences mentioned by the observers at Snag. For days, a small fog or steam patch would appear over the sled dogs, at a height of about 20 feet. It would disappear only in the warm part of the day when the temperature warmed up to −60°F. A chunk of ice was so cold that when brought into a warm room, it took five full minutes before there was any trace of moisture, even when held in the hand.

Blezard recalls antics around the camp during the cold spell:

"We threw a dish of water high into the air, just to see what would happen. Before it hit the ground, it made a hissing noise, froze, and fell as tiny round pellets of ice the size of wheat kernels. Spit also froze before hitting the ground. Ice became so hard the ax rebounded from it. At such temperatures, metal snapped like ice; wood became petrified; and rubber was just like cement. The dogs' leather harness couldn't bend or it would break."

Toole added a few other memories of the cold snap:

"It was unique to see a vapor trail several hundred yards long pursuing one as he moved about outside. Becoming lost was of no concern. As an observer walked along the runway, each breath remained as a tiny, motionless mist behind him at head level. These patches of human breath fog remained in the still air for three to four minutes, before fading away. One observer even found such a trail still marking his path when he returned along the same path 15 minutes later."

Life in the cold had its complications. Surprisingly, heating the log buildings was less of a problem than one might expect. In a memo to his superior, Dr. Tom How, officer-in-charge of the Edmonton forecast office, Toole wrote about the hardships that winter:

> With constant stoking of the furnace the temperature of the barracks remained quite comfortable. The only uncomfortably cool room in the barracks was the common room, this was due to a

large hole, 8 feet by 4 feet, being in the ceiling. The hole was caused by the freezing and bursting of one of the water pipes on December 2nd. Despite promises by the RCAF at Whitehorse that a carpenter was coming up on the first available aircraft to fix the ceiling, the temporary patch, put on by two of the radio personnel and myself, remains....

No provisions have been made for supplying the barracks with water for drinking or washing purposes. This, as you can see, has made it almost impossible for personnel to wash more than once a day and has terminated showers or baths. These unsanitary conditions will very likely continue until such time as the RCAF are prevailed upon to haul water by truck. After seconds outdoors, nose hairs froze rigidly and your eyes tear. Facial hair and glasses become thickly crusted with frozen breath...you had to be careful not to inhale too deeply for fear of freezing or scalding one's lungs. The only other discomfort caused by the cold were numerous cases of beginning frostbite, particularly the familiar "ping" as the tip of one's nose froze. One only had to remain outside for 3 or 4 minutes with face exposed before cheeks, nose and ears were frozen.

In a letter dated August 22, 1995, Toole claimed that the crew had first-class toilet facilities when the indoor plumbing froze up. "A small building was hauled about 100 feet from the barracks. It boasted four seating positions, but best of all, a wood heater that was kept stoked at all times. Talk about comfort!"

During the extreme cold, outdoor chores had to be postponed. The weather staff felt fortunate that observing duties kept them outside only for two minutes every hour. On the other hand, the enlisted men were outside for relatively long bouts, hauling wood to keep the barracks, the garage, and the powerhouse warm. They had to take extra precautions to prevent throat and lung burn from overexertion in the frigid air. Says Blezard, "It was easy to freeze your nose at −70°F without even knowing it was cold. At −30°F you feel it coming."

Beating the "cold blues" was another challenge. Toole busied himself during the cold spell by checking his trap lines; others played poker or hearts, boxed, listened to classical music, read, or talked. And the talk was about the wretched cold. According to Blezard, "When the cold stayed for just a few days, it didn't bother you that much. It was something to talk about, and probably improved the state of mind for a while. But the enduring cold wore you down by sapping your energy."

In the midst of the cold spell, there was no resupply by RCAF planes from Whitehorse for several weeks. Aircraft did not fly when the temperature fell below –50°F. The pilots might get the planes started after several hours, but could not get them warm enough to take off. Furthermore, the pilots were afraid the landing gear would freeze up and crack.

Explained Toole: "In the extreme cold the aircraft would land, but the pilots learned never to use their brakes for long intervals, as they froze up and could not be broken loose until warmed. As a result, the practice was for the aircraft to taxi slowly up to the unloading area, where items were thrown out as the aircraft continued to move. It was extremely cold for the personnel unloading in the prop wash and not the safest, but it worked."

**Extreme cold has played a role in US history: cold weather in New Jersey in December 1776 froze the Delaware River and allowed George Washington and his troops to cross the frozen river and defeat the Hessian troops at Trenton. Then, on January 3, 1777, another cold snap allowed George Washington and his troops to cross the battle line at Princeton, defeat a British regiment, and find safety in the hills to the north near Morristown. Cornwallis and his troops withdrew and American morale was boosted.**

Blezard recalled: "All we ate was fish and bacon and eggs... There was very little meat... We lived mostly on beans for the last five days." So the men were very happy when the cold spell ended and the first airplane (a DC3) could land, bringing meat, vegetables and fruit, a few cases of beer, and a couple of bottles of rye.

Starting machinery was also a chore. And getting an engine started was no guarantee it would continue to run. At that temperature, the oil

**How Cold Does That Wind Chill Feel?**

| Wind speed (mph) | Temperature (°F) | | | | | | | | | | | | | |
|---|---|---|---|---|---|---|---|---|---|---|---|---|---|---|
| | 40 | 35 | 30 | 25 | 20 | 15 | 10 | 5 | 0 | −5 | −10 | −15 | −20 | −25 |
| 5 | 36 | 30 | 25 | 19 | 14 | 8 | 3 | −2 | −8 | −13 | −19 | −24 | −30 | −35 |
| 10 | 26 | 20 | 13 | 7 | 1 | −6 | −12 | −18 | −25 | −31 | −37 | −44 | −50 | −56 |
| 15 | 20 | 13 | 6 | −1 | −7 | −14 | −21 | −28 | −35 | −42 | −49 | −56 | −63 | −70 |
| 20 | 16 | 9 | 2 | −6 | −13 | −20 | −28 | −35 | −42 | −50 | −57 | −64 | −72 | −79 |
| 25 | 13 | 6 | −2 | −9 | −17 | −25 | −32 | −40 | −47 | −55 | −63 | −70 | −78 | −85 |
| 30 | 11 | 4 | −4 | −12 | −20 | −28 | −35 | −43 | −51 | −59 | −66 | −74 | −82 | −90 |
| 35 | 10 | 2 | −6 | −14 | −22 | −30 | −37 | −45 | −53 | −61 | −69 | −77 | −85 | −93 |

Source: National Weather Service

and the transmission fluid coagulated into something approaching a solid. In addition, truck tires could splay open when they hit ruts. But the weather instruments, apart from the thermometers, all seemed to work in the cold.

How does the Snag record stand compared with the rest of North America? It beat the lowest temperature ever recorded in the United States. That was −79.8°F recorded on January 23, 1971, at Prospect Creek, Alaska, a camp along the Alaskan pipeline in the Endicott Mountains. The coldest temperature ever recorded in the lower 48 states was −69.7°F at Rogers Pass, Montana, on January 20, 1954.

How about the rest of the world? Snag still compares well: extremes below −80°F have occurred in only three other places: northeastern Siberia, central Greenland, and Antarctica.

The lowest temperature ever recorded at a weather station was −128.6°F. That was in eastern Antarctica at the Russian scientific station of Vostok on July 21, 1983. Vostok also held the previous world record, −126.4°F on August 24, 1960. This station, which is not staffed year-round, owes its terrible cold to its lofty height of 11,200 feet above sea level.

Where is the lowest temperature where people live year-round? For many years, the coldest settlement was centered a degree of latitude north of the Arctic Circle at Verkhoyansk in northern Siberia. It held the world record for low temperatures at −90.4°F, registered twice in February 1892.

However, some controversy exists about the reliability of the thermometer from which the reading was taken, and there have been several unconfirmed reports of lower temperatures recorded elsewhere. Believed to be an even colder place is Oymakon, a village of about 600 people in a mountain valley 2,300 feet above sea level on the banks of the Indigirka River in northeastern Yakutia. No observations were made in the winter of 1892 at Oymakon, but in later years, it was consistently colder than Verkhoyansk. Oymakon's lowest recorded temperature is −90.4°F in January 1959.

**In Caribou, Maine, elementary school children stay indoors when the wind chill falls below 0°F. Their northern cousins are expected to weather much colder temperatures: in Kodiak, Alaska, children only stay indoors at school when the wind chill falls below -20°F.**

The only other place in the world where temperatures have been colder than −81°F is Greenland. On the permanent ice cap in central Greenland, the lowest official temperature for the Western Hemisphere was recorded on January 9, 1954: −86.8°F at Northice (7,680 feet above sea level), a station established by the British North Greenland Expedition. Since Northice was open for only 20 months, it is probable that lower temperatures might be expected there over a longer period of time.

Will Snag remain North America's cold spot? Only time will tell. But one thing is certain: weather observers will no longer have to mark thermometer sheaths when temperatures fall below −80°F. Now, official alcohol thermometers have markings to −94°F, a thermometer redesign due to the coldest day in North American history.

# The Great Flood of '93

## Moving out of Water's Way

They say you have to go with the flow. Residents of Valmeyer, Illinois, probably wouldn't agree.

The Great Flood of 1993 inundated their small town, some 25 miles south of St. Louis, and left 90 percent of the homes in ruin. The town was immersed under muddy, polluted water from August until October 1993; people had to use boats to get around. It wasn't the first time a deluge of water from the Mississippi River destroyed so much of what these people had built, but many flood-wary residents decided it would be the last. In the wake of the flood, the townspeople voted to rebuild the whole town two miles away and 400 feet higher up on a bluff—away from the flow of the mighty Mississippi.

**Three-quarters of federally declared disasters in the US are floods. On average, floods and flash floods damage $3.5 billion worth of property and take 200 lives in North America each year. About half of flood fatalities are vehicle-related.**

The story of a whole town moving is uncommon, but the story of people locating in the flood plain is not. Valmeyer was founded at the turn of the century, and had grown up in the fertile flood plain of the Mississippi River. It's not surprising that people wanted to live in these areas, because the ebb and flow of the river over centuries made the surrounding soils highly productive, and the river provided nearby transportation for goods. But a big river like the Mississippi gives with one hand and takes away with the other. Its rising waters can be inexorable and devastating.

Valmeyer had flooded before, in 1910, 1943, and 1944. A levee helped stem the tide of the river for many years and many near-floods, but the levees were not enough when the greatest of floods hit in 1993.

Unusually heavy rains drenched the Midwest in the spring and summer of 1993. The previous autumn was also wet, and left the ground

A Manitoba church surrounded by floodwaters. *Tom Hanson/Canapress*

moist. Snowfall in the winter of 1992–93 was heavy, and the snowmelt and spring rains saturated the ground. In 1993, Illinois, Iowa, Minnesota, and North Dakota had their wettest January to August on record. From June to August, a statewide average of 19.67 inches of rain fell in Illinois, and 26.9 inches fell in Iowa. Rainfall in most areas of the Midwest was twice as heavy as normal. The Mississippi, Missouri, and Kansas rivers gathered in the unusually high runoff and swelled to an enormous flood. The high water worked its way gradually downstream toward Valmeyer, but other places flooded first.

**After the 1993 flood, Valmeyer, Illinois, adopted the motto "Rising to New Heights," but they weren't the only ones. A graveyard in Hardin, Missouri, was inundated, and 700 coffins and vaults rose to the surface—some floated as far as 14 miles downstream.**

In Des Moines, Iowa—where the Raccoon River flows into the Des Moines River, which eventually meets the Mississippi—the river crested on July 11 more than 14 feet above flood stage. The city water supply was flooded, and its 250,000 residents went for 12 days without potable water. Nearly a quarter of its residents were also without power, and over 2,000 homes were severely damaged. The whole state of Iowa was declared a disaster area from June to August.

**The Great Flood of 1993 was the worst disaster in the Midwest in the 20th century, with total damages estimated at over $16 billion. The Mississippi River and many of its tributaries flooded after unusually heavy spring and summer rains whipped the Midwest. The flood affected nine states. Fifty-two lives were lost, 75 towns flooded, 70,000 people left homeless, and 60,000 square miles of farmland soaked beyond use. The deluge made 1993 the costliest year for flooding in US history.**

Bridges were washed out. Some 4,000 miles of railway tracks were flooded. From June through August, more than 5,000 barges, laden with goods, were stuck on the Mississippi. In some places, the Mississippi River, normally less than a mile wide, grew to be as wide as seven miles. In total, over half the levees were broken. In St. Paul, where the river peaked on June 26, the downtown airport disappeared under several feet of water. Town after town on the Mississippi, Des Moines, and Missouri rivers got soaked.

Still, downstream of all these disasters, Valmeyer remained dry behind the levees. Although warned of the danger by hydrologists and engineers, many residents of Valmeyer didn't believe they would be flooded.

"We did what everybody else did," said Charlotte Gartzke, who lived with her husband in Valmeyer. "We took our pictures and mementos. But in your heart, you didn't really think it was going to happen. Then it rained and rained and rained and rained."

"The night of the flood," Mayor Dennis Knobloch would say four years later, "I was standing in water, and I was thinking, 'Something's going to stop this water.'"

But nothing could stop it, and nothing did. At 4 a.m. on August 2, the levee at Valmeyer finally broke, and the swollen, muddy Mississippi gushed through homes and businesses, across the baseball field, fairgrounds, and streets. The town had already been evacuated, so no lives were lost there, but it left the town in waterlogged ruins.

"The morning when it broke," said Charlotte Gartzke, "talk about the worst day in your life!" Their old brick home had been under 18 feet of water during the worst of the flood. "We all sat down there and cried. Your house is your home, your friend, your sanctuary. It was like a death."

But the death of Valmeyer eventually had a happy ending.

The aftermath of the flood was tough. Following the ruin of their old town, people lived in temporary trailer-housing provided by the Federal Emergency Management Agency (FEMA). The trailer park, which was eight miles away in Waterloo, became known as "FEMAVille." Many families were cramped. Others simply moved away. Almost none had the option to move back to Valmeyer, since the town lay in soggy ruins. Very little was salvageable.

"We hauled away 12 truckloads of debris," said one resident.

Though Mother Nature showed no mercy in the US Midwest in 1993, the people of Valmeyer proved to be just as tenacious. They decided that the river would never have its way with them again.

The town adopted the motto "Rising to New Heights" and bought 500 acres on a bluff almost overlooking the old town.

The plan to relocate was ambitious. Federal and state funding was about $35 million. The new school alone cost $11 million. There were environmental, political, and legal matters to consider, not to mention sentimental issues. But under the ambitious leadership of Mayor Knobloch, the issues were resolved, and soon new homes, a new school, new churches, and a new community were built on the bluff.

**The Mississippi River flooded dramatically in April 1927—in some places the river grew to 80 miles wide and over 600,000 people were displaced. That flood gave Congress impetus to pass the Flood Control Act of 1928. Levees, reservoirs, and floodways were built, but even the extensive system of flood management along the Mississippi can't prevent the heaviest rains from flooding the low-lying plains.**

## Killer Floods

- Flash floods occur when large amounts of water gush through a narrow passageway like a street or canyon. They can happen in any state, and are usually due to an intense cloudburst or the breakage of a dam or other retaining wall. Just six inches of flowing water can knock a person off his feet. On June 9, 1972, 237 people died in Rapid City, South Dakota, after 14 inches of rain fell in four hours and burst a dam, sending water gushing through the town. In Big Thompson Canyon, Colorado, 156 people died in a flash flood on July 31, 1976, as they prepared to celebrate the state's 100th birthday. Almost a foot of rain fell in a few hours, forcing huge amounts of water to gush through the canyon.
- The deadliest flood in US history was at Johnstown, Pennsylvania, on May 31, 1889. Heavy spring rains and snowmelt burst the South Fork Dam, sending a 35-foot wall of water—about 20 million tons—gushing into Little Conemaugh Valley where Johnstown lay. At least 2,200 people died.
- Flooding in China: China is plagued by some of the heaviest flooding in the world. It holds the record for the most disastrous flood in history. In the autumn of 1887, the 2,900-mile-long Yellow River flooded some 50,000 square miles of farmland. Over a million people died; some reports put the figure as high as six million. It flooded again in 1931; one million people perished because of the flood and ensuing famine and disease. In the summer of 1998, various floods in China claimed 3,656 lives, according to the government, and left 14 million homeless.

Finally, almost two years after the Great Flood sank their town, residents of Valmeyer started to move into their newly built homes in "New Valmeyer." (Fifteen families, whose homes were less than 50 percent damaged, chose not to relocate, and they still live in the flood-prone low-lying land.)

The new town is the largest community in US history to be entirely relocated from a flood plain to higher ground. It's an example that officials at FEMA would like to see followed by others. And as for Charlotte Gartzke, who moved with her husband to the new Valmeyer, almost everything may look different, but there are small bits of reassurance that she is still at home.

"You might be in a different location," said Charlotte Gartzke, "and people you miss don't come back. But it's still Valmeyer. I still have the same zip code."

**The Red River Flood of April 1997 forced 60,000 residents of Grand Forks, North Dakota, and East Grand Forks, Minnesota, from their homes. It was among the largest evacuations of an urban area in the US since the Civil War. On April 21, the river crested at 54.11 feet—22.6 feet above flood stage—and broke the 100-year record. The devastating flood damaged some $2 billion worth of property, making it the costliest US disaster in per capita terms.**

# The Day Niagara Falls Ran Dry

On the night of March 29, 1848, the unthinkable happened. The mighty Niagara Falls eased to a trickle and then fell silent for 30 puzzling hours. It was the only time in recorded history that this wonder of the world had been stilled. So incredible was the event that three decades later eyewitnesses were still being asked to sign declarations swearing that they were there when "the Falls of Niagara ran dry."

Residents first realized that something was wrong when they were awakened by an overpowering, eerie silence. Inspection of the river by torches revealed only a few puddles in the river bed. The next morning some 5,000 sightseers from as far away as Hamilton and Buffalo jammed the roads to Niagara Falls and converged on the river bank to see the phenomenon. The American Falls had slowed to a dribble, the British Channel was drying fast, and the thundering Canadian Horseshoe Falls were stilled. Upstream at Chippewa, the Welland River was reduced to a mere stream. Above the falls, water wheels at flour mills and factories stopped turning as the river level dropped.

**The American Falls were reduced to a mere trickle as ice dams cut off the flow of water in 1883, 1896, 1904, 1909, 1936, and 1947.**

For some, the event was an interesting curiosity. Peering down from the bank, they saw long stretches of drying mud, exposed boulders, and chains of black puddles. Fish and turtles lay floundering in crevices. While thousands stood in disbelief, a few daredevils explored recesses and cavities at the bottom of the dry river gorge never before visible. They picked up bayonets, muskets, swords, gun barrels, tomahawks, and other relics of the War of 1812. Others took the historic opportunity to cross the river above and below the falls—on foot, on horseback, or by horse and buggy. A squadron of US cavalry troops enjoyed the novelty of a ride down the river bed, while some young entrepreneurs parked a cart just above the brink of the Canadian Falls and retrieved huge pine timbers measuring 12 to 18 yards long. Years later, owners of furniture made from those once-submerged timbers delighted in recalling how the wood was obtained.

**On January 9, 1889, the Niagara Suspension Bridge, located just above the Falls, was blown down during a storm described by the press as "one of the greatest storms that has ever passed over any part of Canada."**

Below the falls, the dry river course provided an opportunity to blast out the rocks that had scraped the *Maid of the Mist* since its launch in 1846. As one account stated: "The canyon of the river reverberated to constant blasting as the rocks were blown to pieces and removed with the same ease as if they had been on dry land."

For superstitious people the unusual silence and unexplained phenomenon was a portent of divine wrath or impending doom. As the day wore on, fear and anxiety spread. Thousands attended special church services on both sides of the border. Native people in the area shared in the belief that some disaster was about to happen.

Tension grew until the night of March 31, when a low growl from upstream announced the return of the waters. Suddenly, a wall of water surged down the river and over the falls. The deluge quickly covered the massive boulders at the base of the falls and restored the ever-present Niagara spray. Relieved residents relaxed and returned home to sleep again to the rumble and boom of the falls. The cause of the stoppage, it was discovered later, was an ice jam that had formed on Lake Erie near Buffalo. In an average winter, Lake Erie is almost completely ice-covered. Normally, by the end of March the lake is clear except in the eastern basin near Buffalo where prevailing winds and water currents concentrate drifting ice. Westerly winds blowing down almost 250 miles of open water break the ice into mammoth chunks and remold it into ridges and rafted ice. The thawing process accelerates in late March, especially on sunny days and under the flow of warm, moist, southerly air.

**On January 27, 1938, the Honeymoon Bridge at Niagara Falls collapsed after a massive ice jam pushed it off its abutments. The ice, which piled up in the river after a storm on Lake Erie, also destroyed the *Maid of the Mist* docks and damaged a power plant.**

Contrary to published reports, the winter of 1947–48 was not intensely cold and Lake Erie's ice cover was not thicker than the usual five to 25 inches. In fact, that particular winter was almost 4°F warmer than usual, though the first half of March was unseasonably cold. In late March, several days of stiff easterly winds drove Erie's pack ice up the lake. But on March 29, the winds suddenly reversed direction, coming out of the southwest and west, propelling the vast ice field back down the lake. The ice was melting rapidly as afternoon temperatures climbed to 45°F under clear skies. The combined force of wind, waves, and lake currents jammed hundreds of thousands of tons of ice into a solid dam at the neck of the lake and the river entrance between Fort Erie and Buffalo. Eventually, the ice cut off the water's flow and the basin downstream gradually dried out. The ice dam, however, was under constant assault. The weather continued balmy and nighttime temperatures stayed above freezing. On March 31, the temperature rose to 61°F, the winds continued to shift and strengthen, and that night the Niagara ice wedge dislodged, restoring the river flow.

Will Niagara Falls ever run dry again? Probably not, at least not on its own accord. Since 1964, an ice boom has been positioned at the head of the Niagara River every winter to prevent the formation of ice blockages and safeguard hydroelectric installations.

The falls have been turned off, though. For seven months in 1969, the US Army Corps of Engineers diverted the river to permit repairs to the eroding face of the American Falls. On six other recorded occasions, the American Falls have frozen over completely. February 1947 was especially cold and the channel on the north side of Goat Island, which separates the two falls, became completely blocked with large masses of ice. But not the Canadian Horseshoe Falls. With 10 times the volume of the American Falls, only once has its mighty roar been stilled—on that memorable March night in 1848.

## Ice Storm '98

Ice storms are often winter's worst hazard. More slippery than snow, freezing rain or glaze is tenacious, clinging to every object it touches. A little can be dangerous, a lot can be catastrophic. In early January 1998, a major ice storm hit the northeastern US and Canada. Though half a million New Englanders were without power for some period of time during the storm, Canada suffered the brunt of the storm, and Ice Storm '98 has gone down in Canada's history as its costliest natural disaster with 35 casualties.

These storms are a major hazard in all parts of Canada except the North. They can hit many regions in the US, but are especially common along the eastern seaboard. The severity of ice storms depends largely on the accumulation of ice, the duration of the event, and the location and extent of the area affected. Based on these criteria, Ice

People evade downed tree limbs and utility lines after Ice Storm '98. *Canapress*

Storm '98 was the worst to hit Canada in recent memory. From January 5 to 10, 1998, the total water equivalent of precipitation, comprising mostly freezing rain and ice pellets plus a bit of snow, exceeded 3.3 inches in Ottawa, 2.8 inches in Kingston, 4.3 inches in Cornwall, and 4 inches in Montreal. Previous major ice storms in the region, notably December 1986 in Ottawa and February 1961 in Montreal, deposited between 1.2 and 1.6 inches of ice—about half the thickness from the 1998 storm event!

**Ice storms can also wreak havoc in the South. In February 1987, an ice storm hit the southern US from the Mississippi Valley to the Carolinas. Up to three inches of ice cloaked trees in South Carolina.**

The extent of the area affected by the ice was enormous. Freezing precipitation is often described in weather reports as "a line of" or "spotty occurrences of." At the peak of the storm, the area of freezing precipitation extended from Muskoka and Kitchener in Ontario through eastern Ontario, western Quebec, and the Eastern Townships to the Bay of Fundy coasts of New Brunswick and Nova Scotia. In the United

**What was the extent of damage in the US? The National Climatic Data Center reported that Ice Storm '98 affected upstate New York, northern New Hampshire, Vermont, and much of Maine. Some places were coated in more than three inches of freezing rain, with a radial ice thickness of one inch or more. There were over 500,000 New England customers without power. Eighty percent of Maine's population lost electrical service. Overall damages were estimated at $400 million in the US, with over $300 million of that in Maine.**

States, icing coated northern New York State and parts of New England.

What made the ice storm so unusual, though, was its duration. On average, Ottawa and Montreal receive freezing precipitation 12 to 17 days a year. Each episode generally lasts for only a few hours at a time, for an annual average total of between 45 to 65 hours. During Ice Storm '98, it did not rain continuously; however, freezing rain and drizzle fell for more than 80 hours—again nearly double the normal annual total.

The storm brutalized one of the largest urban areas of North America, leaving more than four million people freezing in the dark for hours, if not days. Sixteen thousand troops were deployed to help with clean-up, evacuation, and security. Millions of people went to shelters or visited family and friends to shower and cook. Some 75,000 miles of power lines and telephone cables were down. About 30,000 wooden utility poles were downed, as well as 1,000 transmission towers. Millions of trees snapped. People said the snapping of the trees sounded like gunshots.

Without question, the storm directly affected more people than any previous weather event in Canadian history. In the third week following the onset of the storm, more than 700,000 people were still without electricity. Had the storm tracked 60 miles farther east or west, the disruptive effect would have been far less crippling.

The damage in eastern Ontario and southern Quebec was so severe that major rebuilding, not repairing, of the electrical grid had to be undertaken. What it took human beings a half century to construct took nature a matter of hours to knock down.

Farmers were especially hard hit. Dairy and hog farmers were left without power, frantically sharing generators to run milking machines and to care for newborn piglets. Many Quebec maple syrup producers, whose trees account for 70 percent of the world supply, were ruined with much of their sugar bush permanently destroyed.

## Setting the Scene

For several days prior to the ice storm, a low-pressure weather system over the Texas Panhandle pumped moist, warm air from the Gulf of Mexico into southern Ontario and Quebec at cloud level. At the same time, over Hudson Bay, a large stationary Arctic high pressure area maintained a northeasterly circulation over central Quebec, draining very cold air into the St. Lawrence and Ottawa river valleys. Unable to dislodge the heavy, cold air in the river valleys, the southerly current overrode the wedge of cold air at the surface, setting the scene for the onset of freezing rain.

**Atlanta was encased in ice during a storm on January 8, 1973. Further north, 300,000 Georgians were left without power for several days when an ice storm left as much as four inches of ice.**

The weather remained unchanged throughout the week, because out in the Atlantic near Bermuda, a large high-pressure system blocked the Gulf storms from following their normal track across the Atlantic and northward to Iceland. Instead, like a boulder in a stream, the high-pressure system diverted the bulk of the moisture farther west along the western flank of the Appalachian Mountains and into Ontario and Quebec where it collided with the cold Arctic air.

Streams of wet, mild air pushed northward throughout the week. Heavy rains caused deadly flooding in some American states and brought a wet January thaw for much of southwestern Ontario before heading into eastern Ontario. Late on January 9, the main weather system broke down, and surface winds veered southwesterly.

El Niño had a part to play in the ice storm. Since early December, a strong subtropical jet stream had flowed from the Pacific Ocean across the southern United States. This flow typically happens during the mature phase of El Niño and results in increased storminess along the Gulf coast of the United States.

## Forecasting a Freezing Rain Storm

Forecasting the occurrence and amount of freezing rain is tricky. While the temperature hovers around the freezing mark, the weather often cannot make up its mind whether to drop liquid or frozen precipitation or some congealed mixture. A one-degree temperature swing on either side of freezing can make all the difference in the type of precipitation that falls.

For freezing precipitation, the atmosphere must be properly layered—a layer of warm air aloft with temperatures above freezing, sandwiched between layers of colder air with temperatures below freezing. Often in winter the warm, moist air overrides the heavier, denser cold air near the surface.

**Nineteen ninety-eight was heralded in with an ice storm, and it also ended with ice. Ice storms all but shut down Williamsburg, Virginia, at Christmastime. About 164,000 people were left without power, and chefs at the Williamsburg Inn cooked Christmas dinner for 300 customers on outdoor barbecue grills. Elsewhere in the South, 45,000 people in Mississippi, 43,000 in Tennessee, and 17,000 in Louisiana were without power because of storms.**

When rain falls, or snow melts while falling through the intermediary warm layer, rain falls into the shallow cold layer hugging the ground. There, with the air temperature below freezing and/or the ground and objects still below freezing, the chilled raindrops freeze but not completely. Instead, they reach the surface as a supercooled liquid (water droplets at a temperature below 32°F) or as a mixture of liquid and ice. Upon striking a colder object, such as the pavement, hydro wires, tree branches, building walls, or cars, the supercooled raindrops spread out and freeze almost immediately, forming a smooth, thin veneer of slick ice. Freezing rain or glaze contains no air bubbles and looks as smooth and clear as glass. If the drops are tiny (less than 0.02 inches in diameter), the precipitation is called freezing drizzle.

# The Year of Weather Discontent

You don't have to look far back in North American history to find one of the craziest weather years on record: it was 1998. With heat, drought, rain, floods, ice storms, hurricanes, and tornadoes, 1998 set weather records all over the United States.

**The American Red Cross reported that 1998 was the most expensive year for its disaster relief program. It spent some $160 million to help victims of 239 natural disasters. What did the Red Cross do? It served over three million meals and cared for more than 200,000 Americans left homeless in disasters. Its most expensive project was Hurricane Georges, which caused a total of $3-$4 billion in damage in Florida, Louisiana, Mississippi, and Alabama.**

Overall, 1998 was the wettest year in the US on record. That's easy for Californians, Midwesterners, and New Englanders to believe, but try telling that to Southerners in July. While others were drenched, millions of Southerners were parched under a scorching summer sun, and saw no rain for weeks.

For perhaps the first time, the bad weather was not the fault of the weatherman. It was the year of "Blame everything on El Niño," even when there was no scientific reason for that blame. El Niño was blamed for everything so far in advance, in fact, that it might have been a disappointment if the weather in 1998 hadn't been so unusual.

Even before the year began, people from coast to coast were talking about El Niño and preparing for its wrath. Congress was warned in October 1997 of the devastating impact that El Niño would have in California. Meanwhile, California residents stocked up on sandbags and designer raincoats in anticipation of forecasted floods. El Niño was both accused of and credited for everything good and bad about 1998.

Unfortunately, the weather lived up to the warnings. It was the deadliest Atlantic hurricane season in two centuries, and the deadliest tornado year in the US in 24 years. A merciless heat wave throughout the South took at least 200 lives and destroyed some $6 billion in crops. Four major US weather disasters caused at least a billion dollars damage each.

**The deadliest disaster in the US in 1998 was the heat wave that dragged the South through a scorching summer. Over 200 people died.**

The year began with a chilly bang. Not even a week into the New Year, severe ice storms shrink-wrapped the Northeast and Canada in a crystal shroud, and left half a million New Englanders without power. Canada was hardest hit, with four million people losing power. In Maine, 80 percent of households were without power at one point or another during the storm. Without electricity, farmers couldn't milk cows, and even when the cows could be milked, by hand or with the help of a generator, the lack of refrigeration soured the milk. In some areas, up to an inch of heavy, solid ice encased trees, power lines, houses, and cars.

The National Guard came out to help, and federal disasters were declared in Maine, New Hampshire, Vermont, and New York. Estimates of damage in the US were between $400 million and $500 million. Sixteen people in the US and 28 in Canada died. A second storm hit New England two weeks later, knocking power out in thousands more homes.

**It was an unusually bad year for tornadoes in the US. Although over the past few decades, tornado deaths have been generally decreasing due to better methods of forecasting and warning, twisters in 1998 claimed 124 lives—the highest toll since 1974.**

While New Englanders were trying to keep warm and waiting for their power to come back on, record precipitation was falling in Louisiana, Alabama, Virginia, and Tennessee. Snow up to 18 inches deep blanketed parts of Tennessee. A dozen Americans died in floods in Tennessee, the Carolinas, Kentucky, and Alabama.

In February, heavy El Niño rains, floods, and mudslides gouged the California coastline and caused over half a billion dollars in damage. It was the wettest February on record in California. Many towns and cities set records for total rainfall. A eucalyptus tree was blown onto a sport utility vehicle in a Los Angeles suburb, killing

Millions of dollars worth of damage was inflicted on Tegucigalpa, Honduras, by Hurricane Mitch in 1998. *Canapress*

two people. Two Highway Patrol officers died near Santa Maria when their car was washed into a river. Roads were gushed out, major highways were closed because of flooding, huge rocks and mudslides careened onto the Pacific Coast Highway, 2,000 Santa Paula residents were evacuated from their homes, and Santa Barbarans waded through waist-high water to get to shopping centers. Near San Francisco at Pacifica, ocean cliffs crumbled under the power of the pounding surf and left houses dangling. Residents of La Conchita shoveled mud from their driveways and garages. Train service from Los Angeles to Seattle was halted when a trestle was wrecked in a storm-driven surge. Seventeen people died in various weather-related accidents in California.

**The costliest disasters in the US in 1998 were the southern heat wave (over $6 billion in damages); Hurricane Georges ($3 billion to $4 billion); and Hurricane Bonnie (about $1 billion).**

On the other side of the country, thunderstorms and tornadoes raged through the darkness in central Florida on the night of February 22 and into the morning of February 23. The death toll reached 42—the deadliest

Florida tornado in history. Eight hundred homes were ruined, and 700 others badly damaged. Over 100,000 residents were left without power. Disasters were declared in four-fifths of Florida's counties, and when the shards had settled, $100 million worth of damage was left.

And snowstorms, heavy rains, and floods—not to mention tornadoes—kept coming. In early March, after heavy rainfall in Alabama, a levee failed and water surged into the town of Elba, killing four people and damaging over 1,000 homes. Blizzards in Indiana and Illinois left hundreds of thousands without power, and two Indiana residents died from heart attacks while shoveling snow. Snow in Des Moines, Iowa, drifted 12 feet high.

**The US wasn't the only country that had a bad weather year. Other major disasters of 1998: killer Hurricane Mitch in Central America, extensive flooding and mudslides in Peru, severe drought in Australia, record flooding in Bangladesh—some two-thirds of the country was flooded—and the flooding of China's Yangtze River. Chinese officials estimate that the Yangtze River flood left 3,000 people dead and four million homeless. Damage was estimated at more than $20 billion.**

A tornado tore through Hall County, Georgia, in the early morning of March 20 and killed 12 people. Later that day, two North Carolinians died in another tornado. On April 8, an outbreak of powerful tornadoes ripped through Alabama—including a rare, enormously powerful tornado rated at the highest level on the tornado-watchers' scale—F5. The tornadoes killed 34 people and damaged about $300 million worth of property. A tornado lashed Nashville on April 16 and injured over 100 people. Other tornadoes tore through Tennessee, Arkansas, and Kentucky on the same day, and a total of 12 people died.

On May 30, South Dakota suffered its first tornado deaths in 28 years. Six people were killed and most of the buildings in the small town of Spencer were ruined when an F4 tornado blasted through. Even though the town was destroyed, the people of Spencer were tenacious. The bank was wrecked, so a teller set up a rudimentary system in her own home and helped customers there.

In the Midwest, it was the wettest June in 70 years. Rainfall in some areas was twice as heavy as normal. New England was also dripping wet: spring in Rhode Island and Massachusetts was the soggiest on record. Cities in Tennessee, Kansas, and West Virginia set records for high rainfall. By early July, 17 states had rivers which were near or above flood stage. The National Oceanic and Atmospheric Association (NOAA) reported that 80 people suffered flood-related deaths in the US from January to July.

**In the Sudan, Africa's largest country, people battled the worst flooding of the Nile in half a century. Over 200,000 people were left homeless.**

Severe summer storms in early July ravaged the country from Maine to Iowa. Five federal disasters were declared in two days. Later in July, severe storms hit Michigan, Tennessee, and Wisconsin. In August, the town of Marion, Indiana, was drenched with an incredible six inches of rain in just six hours.

But spring and summer weren't wet and stormy everywhere in the US. While rain soaked the northeast and some areas of the Midwest, the South was baked in heat and dryness. Drought in the South and Southeast wiped out many types of crops and caused over $6 billion in damage. Losses in Texas and Oklahoma were the heaviest. Florida, Georgia, and the Carolinas also suffered. The Climate Prediction Center reported that April to June 1998 was the driest spring on record in Florida, Texas, Louisiana, and New Mexico. In Miami, Tampa, Daytona Beach, and St. Petersburg, June 1998 was the hottest—or tied for the hottest—month on record. Texas, Louisiana, and Arkansas also broke heat records.

**In India, some of the heaviest rains in decades led to extensive flooding and landslides. Over 2,000 people died in Uttar Pradesh, and wildlife fled in droves, as though clamoring for a space on Noah's Ark. In Assam, five million people lost their homes; much of Kariranga National Park was flooded.**

Nurtured by broiling heat and extreme aridity, fires gobbled up huge chunks of Florida forests in May and June. The fires were especially bad because heavy February rains had caused vegetation to flourish, providing incendiary fuel once the weather grew dry. An estimated $300 million in timber went up in smoke. The fires ate about 485,000 acres and cost some $100 million to fight. Firefighters from as far away as Montana and Oregon were flown in to help battle the blaze.

Texas wilted in its hottest and driest spring and summer. Dallas coped with 100°F heat for almost a straight month, and there were 38 days in which the mercury didn't drop below 80°F. College Station received just 14 percent of its normal precipitation from April to July. Houston and Lubbock withered in their driest April to July periods on record. July was the hottest month on record for Del Rio, Austin, and San Antonio.

**A cold snap in Europe in November 1998 killed at least 130 people in Romania, Poland and Bulgaria. It was so cold in Paris that the Eiffel Tower fountains froze, and a metro station was opened as a shelter for the homeless.**

The human toll was terrible. By the time the heat wave ended, at least 200 deaths had been blamed on heat in the US. Forty-three Mexican immigrants seeking a better life in the US died from the heat. CNN reported that Mexican TV stations were broadcasting a warning to potential border crossers: "Keep Out, Keep Alive."

The impact of the drought on agriculture was grim. At least 40 percent of Texas' corn, sorghum, and cotton crops were assessed to be in poor or very poor condition. Pollination was impeded by the weather. Many chickens died, unable to stand the heat, though in Oklahoma poultry farmers sprayed their lucky fowl with mist to help them stay cool. Citrus trees slumped in the heat. Hay production was decimated, and the little hay that was available was very expensive.

Grasshoppers prospered because there were no early rains, which typically suppress population explosions. They gorged themselves on cotton and what little grass was left. There were food shortages for Texas' five million cattle. It was several years before ranchers recovered. Thousands of agricultural jobs were lost. Total losses in agricultural products in Texas were over $2 billion, with the statewide economic impact estimated at some $5.8 billion. Timber losses were significant, too, with an estimated loss of almost $342 million. Across the state line, in Oklahoma, agricultural losses were estimated at $2 billion. The heavy

August rain and wind further damaged the already-suffering cotton crops. An economist reported that 1998 yielded one of the worst cotton crops ever in Texas.

**The Worldwatch Institute released a study in November 1998 which concluded that the worldwide damage caused by natural disasters in 1998 was a record $89 billion—more than the entire losses sustained throughout the 1980s.**

The crazy weather year sent graphs bouncing in dramatically different directions. Del Rio, Texas, was broiled from May to July, but in early August it was drenched by the heavy rains of Tropical Storm Charley. More rain fell there in 24 hours than the city normally gets in an entire year. The Rio Grande surged to record levels and water from San Felipe Creek gushed through the streets of Del Rio's poor neighborhoods, taking homes, cars, and asphalt with it. There were nine deaths due to flash floods in Del Rio, and four Mexican citizens died when their pickup was washed from the highway into a river in nearby Real County, Texas.

And the flooding wasn't over for the Lone Star State. Heavy rains in October brought more floods to Texas and claimed 22 more lives around San Antonio. By October 21, a quarter of the whole state was under water. The Guadalupe River swelled to a width of up to six miles. Floods in Victoria and Cuero were considered the worst in recorded history.

But for all the extremes of 1998—the flooding, the heat, and the tornadoes—it was the hurricanes that took the most lives and property.

**Worldwide, some 32,000 people died in 1998 as a result of natural disasters.**

The hurricanes and tropical storms began arriving in August, and had friendly names like Bonnie, Charley, Earl, Frances, Georges, Hermine, and Mitch. But by the time the last of these uninvited guests dissipated and drifted back to sea, they had taken a devastating toll on the Americas. They had evicted residents from their homes, annihilated property, obliterated vital crops, and killed thousands and thousands of people with their strong winds, heavy rainfall, surges, and floods.

Ten of the 14 tropical storms that brewed in the Atlantic developed into hurricanes, three of which were classified as major. Seven storms slammed the US—twice as many as the average, and the most to hit the US since 1985. In September, four hurricanes were simultaneously assailing the Atlantic—the first time that's happened in over a hundred years.

Overall, 1998 was the deadliest Atlantic hurricane season since 1780. Most of the people killed were in Central America, and to a lesser degree, the Caribbean. The National Hurricane Center estimates 11,629 people died in tropical cyclones in 1998. Almost all of these people perished in Mitch, which slammed Central America on October 29 and took a deadly path across Honduras and Guatemala before turning north to hit Florida as a downgraded tropical storm on November 4–5. Mitch finally drifted back to sea on November 5. Damage in the US in 1998 due to hurricanes and tropical storms was $6.5 billion. In Central America, damage was estimated to be at least $5 billion, though the true extent of human and economic losses there may never be known. Mitch, whose winds reached 180 miles an hour, was one of the strongest Atlantic storms ever recorded. Its floodwaters and mudslides swallowed entire towns in Honduras and Nicaragua. In one swoop, the infamous storm obliterated years of economic and social progress from the map of Central America. Though the Atlantic hurricane season officially ended on November 30, people were picking up the broken pieces for many years to come.

And it wasn't over yet. There were repeated cold snaps and heavy snow throughout December in many parts of the country. In terms of cold, Montana was among the hardest hit. In Butte, it was −39°F on December 21. On New Year's Eve, parts of New England and Michigan reported wind chills as low as −60°F. Meanwhile, storms and avalanche warnings left some 4,000 travelers stranded in Washington state.

Chicago welcomed the New Year with its worst blizzard in 20 years. O'Hare Airport was closed. Flights all across the Midwest were canceled. Stranded travelers trying to return home after the holidays were no doubt hoping that this was not an indication of what the weather would be like in 1999!

# Weather Across North America

# Blame It on El Niño!

He was the most talked-about child of the year. He was blamed for everything from outbreaks of malaria and dengue fever in Africa, to high ice-cream prices and a frenzy of grasshoppers in the US, from drought and fires in Australia, deadly flooding in China, and the failure of coffee crops in Colombia, to a new lake in Peru.

Who is this child?

For centuries, Ecuadorean and Peruvian fishermen used the term "El Niño" (Spanish for the "boy child" or, by implication, the Christ-child) to describe a weak, warm ocean current that arrived off the coast around Christmas. In some years, though, an unexplained reversal in atmospheric circulation over the South Pacific and Indian Oceans causes the current to warm even more—from 4°F to 9°F above the average maximum sea-surface temperature of 82°F—and expand to cover an area up to three times as large as the US. The effect of this pronounced warming on the world's weather is so dramatic that today the name El Niño is used to describe such exceptional events.

**Especially strong El Niños occurred in 1891, 1899–1900, 1925–26, 1931, 1941–42, 1957–58, 1965, 1972–73, 1976–77, 1982–83, 1986, and 1992. The El Niño of 1997–98 was among the strongest.**

Though it was known to Spanish explorers of Latin America nearly 400 years ago, El Niño was not recognized as part of a global system that caused drought in some regions and fierce rain in others until the middle of the 20th century. Today, scientists still don't know exactly what triggers the process that leads to El Niño. There have been many suggestions: stiffening trade winds and water accumulating in the western Pacific, changes in the salinity of the oceans, and lava gushing out of undersea volcanoes. Though nobody is certain what causes it, much insight into the events leading up to El Niño has been gained in recent years.

The changes in ocean circulation that produce El Niño have been linked to what is called the Southern Oscillation, a huge seesaw pattern of atmospheric pressure between the eastern and western tropical Pacific.

Most of the time, the equatorial trade winds blow west, from a persistent high-pressure system over the southeastern tropical Pacific near Tahiti, toward an equally tenacious low-pressure system parked over Indonesia and northern Australia. This westward flow of air drags warm surface water westward, raising sea levels off the coasts of Indonesia and northern Australia by 12 to 28 inches, and turning the western Pacific into a big storehouse of energy. Meanwhile, on the other side of the Pacific, offshore winds along the South American coast strip away the shallow surface waters, causing an upwelling of cold, nutrient-rich water.

**Measuring Sea Surface Temperature: The National Oceanic and Atmospheric Association uses both its own satellites and NASA's satellites to collect data. It also has a set of 70 buoys in the equatorial band of the Pacific Ocean to measure changes in water temperature, currents, and winds.**

But during an El Niño year, the equatorial trade winds mysteriously slacken or change direction, flowing east instead of west. No longer supported by the trade winds, the giant underwater wave that has accumulated in the western Pacific sloshes back toward the coast of South America, much like the water in a bathtub. This warm eastward countercurrent becomes stronger as it makes its way back across the Pacific. On its 2½ month journey, it also becomes warmer under the hot tropical sun. Eventually, it reaches its destination and overrides the normal upwelling of cold water off the coast of Peru and Ecuador.

The changes in oceanic and atmospheric circulation associated with El Niño cause atypical weather patterns around the world. Rising heat and moisture from the ocean off Peru and Ecuador provide the raw energy for more frequent storms and torrential rainfalls over these normally dry regions. In the northern US, the additional heat strengthens and changes the path of the jet stream, which is the high-altitude, fast-moving river of air that steers weather systems around the world. A diverted jet stream can wreak havoc with the weather wherever it goes.

What happens to North American weather in the weeks following the onset of El Niño largely depends on whether the jet stream remains a single stream or splits in two. A single jet stream, curving north over British Columbia then plunging south through the center of the continent, brings colder temperatures to the Great Lakes region and eastern North America. At the same time, a high-pressure system stalls over the Rocky Mountains, preventing moist Pacific air from moving inland. Mild, dry weather then dominates Western Canada and Northwestern US.

**El Niño is technically classified as a "transitory irregularity in the global ocean-atmospheric system."**

If the jet stream splits, the northern branch tends to create storms in the Gulf of Alaska and warm temperatures in the western US. Its southern branch brings storms to California, Texas, and Florida, before moving up the east coast of North America.

El Niño events are far from regular. They occur, on average, every three to five years, but the interval can vary from two to 10 years. Typically, El Niño lasts from 12 to 18 months, though some expire within a few months.

**El Niño can set in at almost any time of the year; however, it usually occurs early in the calendar year. By about August, it is generally possible to determine whether El Niño is underway.**

Though no two El Niños or their effects are exactly alike, what is good news is that El Niño and its associated weather may be predictable a season or even a year in advance. The first step toward such long-range forecasting is deciding whether El Niño will continue to exert its global influence on the weather. Some work is already under way, using satellites and shipboard instruments to monitor the key early warning signs of El Niño: warming of the surface water off the coast of Peru, shifts of the trade winds, and differences in atmospheric pressure between Australia and the eastern Pacific. Whether or not it is possible to forecast the emergence of El Niño, it may be possible to get at least a few months' warning of an increased risk of weather-related disasters. But long-range forecasting is not easy and, so far, predictions are not reliable enough to be useful.

However, much of the doom that scientists predicted for 1997–98 came true. It was the year of "Blame it on El Niño."

## The El Niño of 1997-98

El Niño can have good and bad effects. Whether it's because we are pessimists or because the benefits of El Niño were so few, everyone seems to talk about the cost of El Niño, and the drastic toll it has had on human lives, crops, animals, and economies.

**Scientists at the University of Guelph in Ontario, Canada, believe that the teeth of female dusky dolphins can be used to date the occurrence of El Niños. Dolphin teeth accumulate a new layer of dentine each year. In female duskies, the layer for an El Niño year is unusually thin, probably reflecting a shortage in food supplies.**

Americans were bracing for El Niño even before he arrived. He was the most talked-about weather phenomenon, and by the time his sister, La Niña, moved him out of the limelight, El Niño had lived up to his reputation.

The World Meteorological Organization said that the 1997–98 El Niño was the strongest of the century. Judge for yourself. Here are some of the things that El Niño has been blamed for:

- It doubled the number of asthma attacks, and caused an epidemic of migraines among headache sufferers in Calgary, Alberta
- It caused rare "tiger belly" whale beachings in southern China
- It encouraged 1.5 million pink flamingos to return to the emerald-green waters of Kenya's Lake Nakuru in the Rift Valley
- It caused a shortage of gourmet coffee beans in Colombia
- It knocked "The Rosie O'Donnell Show" off the air in a dozen East Coast markets when the satellite feed was interrupted
- It brought on an epidemic of skin rashes, itching, burning sensation, swelling, and scratching in Kenya due to the emergence of Nairobi flies
- It forced the price of soybeans and tofu to skyrocket
- It infected college students in the American Northeast with

premature spring fever, making students subsequently flunk mid-term exams

- It snowed in Guadalajara, Mexico, for the first time since 1881
- It decimated the coyote population in Alberta because there were fewer starving and injured deer for them to track down
- It destroyed tens of thousands of sea turtle eggs at Playa La Flor, Nicaragua
- It confused Pacific sockeye salmon, which began showing up ready to spawn in streams where they weren't supposed to be
- It raised the passions of people in Guyana, resulting in a rise in domestic violence
- It messed up precise astronomical observations because increased water content in the atmosphere alters scientists' data by as much as two percent
- It was responsible for malaria, cholera, and E. coli in parts of Africa hit by floods, and dengue fever and hantavirus in South America. In fact, one study performed by the London School of Hygiene and Tropical Medicine concluded that for every degree Celsius (about 1.8°F) that the water in the Eastern Tropical Pacific warms above normal, some 100 million people will need help because of corresponding natural disasters
- It killed off huge numbers of boobies (seabirds) and led to the failure in reproduction of albatrosses, penguins, and cormorants in the Galapagos
- An Antarctic snow bird called the South Polar skua spent several weeks on a warm beach in Florida. It was the first time the skua had been sighted there. Blame it on El Niño? Maybe, but the lost bird was likely blown off course by Hurricane Mitch
- It bloated grapes, resulting in watery, insipid wine
- It caused a marked increase in the number of cases of diarrhea in Peru by polluting water supplies with heavy flooding
- It made 1998 the overall worst year for American farmers since the notorious dust bowl of the 1930s. No rain and lots of grasshoppers in Oklahoma and Texas led to little food for cattle. Skinny cattle had to be auctioned, and cattle losses were at least $44 million

- It decimated cotton production in the US—losses were over half a billion dollars
- It caused an outbreak of wheat scab fungus—caused by humidity and cool temperatures—in the Dakotas
- Its heavy rains in California allowed fire ants to flourish. The ants, who love warm, moist climates, sting people, contaminate water systems, and infest electrical equipment
- Meanwhile, it killed the fire ants in Texas and Oklahoma, where they were needed. Fire ants eat grasshoppers, but it was so dry in the Southeast that the fire ants died. Grasshoppers flourished, dining on cotton and grass, and the cows went hungry
- Its hot, dry weather in Texas and Oklahoma prohibited the growth of various bacteria and fungi that kill young nymphs, another reason grasshoppers had a great year
- It caused incredible fires in Indonesia in October 1997. Continued dry conditions left Southeast Asia quilted in a huge cloud of smoke—some 300 miles long and 185 miles wide. Around 40,000 Indonesians battled eye and respiratory infections. Two planes crashed because of the low visibility
- It caused huge chunks of the Amazon rain forest to go up in smoke
- It rained so hard in Peru that torrential mudslides ruined five percent of the country's roads. Thirty bridges were washed out. A quarter of a million people were left homeless. Diseases like cholera and mosquito-borne malaria broke out. Between $800 million and $1.8 billion in damages were sustained
- It caused a drought in Hawaii and destroyed crops and flowers
- It caused starvation among seals in Southern California because

**The earliest documents referring to El Niño date back to 1795 and are credited to a Captain Colonet. El Niño was first recognized as part of a giant ocean-atmosphere circulation pattern in the 1920s by Sir Gilbert Walker. The pattern is known as the Walker Circulation or the Southern Oscillation.**

the warm water pushed away fish and squid that the seals typically eat

- It clogged traffic in the Panama Canal. Water levels were so low that boat traffic was restricted
- It caused the first green Christmas ever in Saskatoon, Saskatchewan
- It flooded more than 22,000 homes in southern Russia, after heavy snow melted
- It harmed all kinds of crops in California. The cool, rainy spring hurt cherries, almonds, asparagus, watermelon, artichokes, and tomatoes
- It irritated Caesar-salad lovers when it caused a significant shortage of romaine lettuce in the US. California grows about four-fifths of the country's romaine, and when heavy rains there ruined the crops, the price of romaine quintupled
- It took a toll on the quail population in Texas. The drought left the birds without good nesting locations, not enough moisture for egg incubation, and not enough food for the dozen or so young that a mother quail typically has
- It caused higher-than-normal levels of Lyme disease on the East Coast. The rain made grass lush and acorns plentiful, which was good for rodents and the parasitic ticks that carry Lyme disease
- It formed a new lake in the deserts of Peru
- It sent anchovies off the coast of Peru swimming south toward Chile. The anchovy industry—worth about a billion dollars in Peru—was destroyed. Anchovy prices skyrocketed
- It drowned 140,000 cattle in one province of Argentina. Its heavy rains flooded parts of Paraguay, Uruguay, and Argentina when the Parana River swelled beyond capacity
- It forced up the price of ice cream and butter all over the US because heavy rains in California—which has 20 percent of the country's dairy cows—reduced milk production
- It provided ideal breeding grounds for ants, earwigs, and termites in the Northeast
- It upset Florida golf-course owners, who lost money when rainy days kept clients indoors

- And the reason Barbra Streisand didn't win an Academy Award? Billy Crystal said it was El Niño

Wasn't there anything good about El Niño? In 1998, as Americans were soaked, scorched, and irritated, it didn't seem like there were many benefits to El Niño, but some companies cashed in on the hysteria. In general, anyone who sold goods or performed services that helped people cope with the wrath of the wacky weather profited from El Niño.

- In the pre-El Niño frenzy, a Southern California sandbag company quintupled its sales
- The vice president of an Irvine data retrieval company told the *LA Times*, "Disasters always mean good business for us"
- Roofing repair companies did a booming business even before El Niño hit Southern California, as people geared up for the predictions
- Insurance agents reveled in increased sales of flood insurance
- The *LA Times* reported that trendy raincoats were a hot seller in the fall of 1997, as were expensive designer black patent rain boots
- In Santa Monica, unusual and exotic fish—yellowtail, bonita, and bluefin—arrived from Mexico in the summer of 1997. Locals claimed it was the best fishing season in a generation
- One study concluded that El Niño cleansed the air around Southern California in the fall of 1997 because it brought low pressure to the California coast. That brought moisture, clouds, and cool air, which reduced smog. One meteorologist said that 1997 would be the cleanest year on record
- El Niño causes some plants to flourish, according to a study performed at the National Center for Atmospheric Research. The result? Though initially El Niño causes more $CO_2$ to be released into the air, eventually (within two years), the flush of plants absorbs $CO_2$. The net impact, though, is unclear
- New Mexico ski-resort owners delighted in the heavier-than-average snowfalls
- El Niño reduced Atlantic hurricanes in the 1997 season because

it trimmed off the tops of tropical storms that might have developed into hurricanes

## La Niña

About every four to five years, a pool of cooler-than-normal water—as much as 36°F below the average maximum sea-surface temperature of 82.4°F—replaces the warm El Niño current off the western coast of South America. The conditions associated with this cooler water are called "La Niña." La Niña typically lasts nine to 12 months, though it may continue for as long as 24 months.

The effects of La Niña contrast sharply with those of El Niño. La Niña generally brings wetter monsoons to India, flooding to Bangladesh, and heavy rains to northern Australia, southern Africa, Hawaii, and parts of Indonesia. In the US, it usually brings colder winters to the Northwest and Alaska, as well as across the Great Lakes and the Northeast. It generally delivers drier, warmer weather to the Southeast in the winter, and drier weather to the Southwest from the summer through the winter.

The 1997–98 El Niño is was followed by a La Niña. The National Oceanic and Atmospheric Association classified this La Niña as moderate. However, one of La Niña's typical effects, according to Dr. William Gray of Colorado State University, is increased hurricane activity in the Atlantic.

But what Atlantic hurricanes did in the autumn of 1998 was anything but moderate.

# Weathering Heights— A Global Warning?

An increasing number of climatologists believe that unusual and record-breaking weather events are becoming more common and may even be the "fingerprints" of global warming.

On average, the world is 0.9°F warmer today than it was a century ago, with more than half of the increase coming in the past 20 years. Globally, 10 of the warmest years on record have occurred since 1983, with 1998 being the warmest. A warmer-than-normal year is not necessarily proof of climate change; for perspective, longer-term trends worldwide from a decade or more must be assessed. In December 1995, an international panel of scientists did just that and concluded that "global mean temperature changes over the last century are unlikely to be entirely due to natural causes... Evidence suggests a discernible human influence on global climate."

**The Earth's 10 warmest years since 1881 have been: 1983, 1987, 1988, 1990, 1991, 1994, 1995, 1996, 1997, and 1998.**

Before 1998 came along, 1997 was the warmest year of the century, based on global land and ocean surface temperatures. But 1998 was yet another record-breaking year. The extent to which temperature records were broken was most significant in 1998. Each month, from May 1997 to October 1998, global near-surface land and ocean temperatures were record warm. January and February 1998 were the warmest and wettest on record in North America. On average, the global temperatures recorded from January to September 1998 were 1.25°F greater than the average from 1880–1997. The warming continued, and September 1998 was the warmest September on record in the US, warmer even than the dust bowl during the 1930s. Of the 10 warmest years in the past 140 years, seven were in the 1990s and three were in the 1980s; 1998 was the 20th

**North America's average temperature from January to September 1998 was 1.25°F greater than the average from 1880–1997.**

consecutive year with above-normal global surface temperatures.

There is more to climate change than higher temperatures. Computer models of the atmosphere predict that global warming would also bring more volatile weather and related disasters. Climatologists and others who monitor trends and extremes suggest this is already happening, though a correlation between global warming and specific events is difficult to confirm. Nevertheless, over the past decade, more areas of North America have suffered through either extreme drought or extreme wetness than during any other period this century. Heavier precipitation events are on the rise in North America. The summer of 1998 was especially bizarre: Florida had an unusually wet spring, followed by a record hot and dry summer. In Texas, severe drought and heat crippled agriculture and killed people; then, in a cruel reversal, Mother Nature deluged Texas with torrential rains in August.

**The average global surface temperature of the Earth is 59°F; without the greenhouse effect it would be a chilly −0.4°F.**

**On June 19, 2000, the residents of Barrow, Alaska, experienced their very first thunderstorm. Thunderstorms are rare on the Arctic coast because the adjacent ocean is covered in ice and the Brooks Range prevents the milder air from the south from reaching the more northern parts of Alaska. The storm dropped 0.16 inches of rain on Barrow, a town that usually averages only about four inches of rain per year.**

And people around the globe had to cope with plenty of weird weather throughout 1998. It seems as though no region was spared, from drought in Australia and Indonesia, to extreme flooding in China and Bangladesh, from a chilling and deadly cold snap in the autumn in Europe, and the deadliest hurricane to hit the Atlantic in over 200 years, to devastating flooding in Peru and many other parts of South America. Overall, in 1998, an estimated 56 countries suffered extreme floods, while 45 faced severe drought.

What does all this mean? It would be fiction to interpret every weather catastrophe, every few weeks with heat and rain, or every "storm of the century" as clear evidence that the Earth is warming up. Just as any heat wave does not prove the theory of global warming, any lengthy cold snap doesn't mean that the threat of the greenhouse effect is a lot of "hot air." The Earth's atmosphere system is complex and variable.

**Industrialized countries produced 74 percent of the carbon dioxide going into the atmosphere in 1985. The US consumes 15 times more energy per capita than does a typical developing country. The US has about four percent of the world's population, but produces over 20 percent of the world's greenhouse gases.**

Although it's tempting to connect weather surprises with humankind's increased burning of fossil fuels, weather extremes and "yo-yo" temperature swings may have nothing to do with global warming. Instead, they may only be random meteorological aberrations—all part of the natural year-to-year variations in weather. Recent freakish weather is just as likely to be blamed on curvier jet streams, more frequent and longer El Niños, or changes in sunspots.

Most scientists do agree that small changes in temperature trends will lead to larger changes in the frequency of extreme and unusual weather. Their understanding comes largely from the performance of highly sophisticated computer models. Much of what we have observed in temperature trends in the past 100 years is consistent with model projections of the combined effects of greenhouse gases and aerosols on world climates. The global models were impressive in demonstrating the cooling trend following the eruption of Mount Pinatubo in the Philippines and the warm atmosphere resurgence after the volcano's effect had abated.

**The difference in global temperature between now and the last ice age is only 8.1°F.**

However, climate models can as yet only give us some very broad indications of how weather extremes may change. For instance, they suggest that:

- Even a small warming could lead to large increases in the frequency and intensity of droughts, floods, and forest fires
- Hurricanes in a greenhouse-warmed world may be generally weaker, but once formed, they would have the potential to stay intact longer as they moved farther north

- Warmer weather would likely bring more destructive tornadoes, severe hailstorms, intense thunderstorms, and windstorms

Over the past decade or so, observational evidence has suggested that the world's weather is not just warming up, but also becoming more volatile and variable. For example, recent studies reveal that:

- Between 1991 and 1994, there were three El Niños in four years—unprecedented in the past 200 years. In recent years, these warming episodes have been more frequent and more intense, including two "El Niños of the century" separated by only 15 years
- In the past 10 years, the American Southeast has experienced its worst drought in 300 years, and the Midwest has had its worst flood in history
- Since 1980, Canada has suffered five of its worst forest-fire years in history
- The worst bush fires in 200 years ravaged eastern Australia at the close of 1993
- Insurance claims due to natural disasters in the 1980s were tenfold greater than in the 1960s and even higher in 1991 and 1992; storms have accounted for 88 percent of the disasters in the past decade. Weather-related disasters have cost the planet an average of $1 billion each week over the past few years
- Worldwide catastrophic windstorms numbered 29 during the 1980s, 14 in the 1970s, and eight in the 1960s. In 1992–93 alone, hurricanes, floods, and blizzards in the United States inflicted more than $65 billion in property damages and losses to its economy

**The frozen continent of Antarctica has warmed a whopping 4.5°F in the past 50 years. In the 1970s, glaciologists predicted that melting of the Antarctic ice shelf would be a clear signal of global warming. In 1995, an enormous chunk of ice, 23 by 48 miles in size, broke loose from the western Antarctic ice shelf and floated out to sea. The iceberg was 656 feet thick.**

- According to estimates by the Worldwatch Institute, losses from weather-related disasters for the first 11 months of 1998 were 48 percent higher than the previous one-year record of more than $93 billion in 1996. The year's damage was for ahead of the total losses for the entire decade of the 1980s

**If all of Greenland's icecap melted, sea levels would rise by almost 25 feet; if the Antarctic icecap melted, global sea levels would rise by 213 feet.**

It is too soon to link definitively all this wild and extreme weather to global warming. But when many unusual weather events happen at the same time—and year after year—it is hard to attribute them all to natural variability.

"Increasingly the gut feeling is that they are linked to global warning," says a government climate advisor. So far, science has not proved all the gut feelings, but the intensity of that research, too, is heating up.

# Hurricanes: The Storms with Names

"It looked like Beirut," said the mayor of Homestead, Florida, 30 miles south of Miami. "There were vehicles driving with no windshields. There was looting. The electricity was totally gone. We were approaching anarchy. We had one phone line. Alex [the city manager] called the President on a Wednesday evening. We said we needed troops."

***Cyclone* was the name given to intense circular storms by Captain H. Piddington in 1848. Today, cyclone refers to the family of storms that includes hurricanes, typhoons, tornadoes, and depressions.**

The troops arrived on Thursday. Eventually there would be 16,000 of them.

It sounds like a war zone, but the disaster that hit southern Florida during the early hours of Monday, August 24, 1992, was not her man-made. Nature did it.

**Several derivatives of the word *hurricane* exist. One version originates with the natives of the West Indies, based on the word *huracan*, referring to a "great wind" or evil spirit. Another possibility is the Guatemalan Indian god for stormy weather, *Hunraken*.**

The hurricane named Andrew had just lashed the Sunshine State, and in the span of only a few hours, Andrew trashed everything it touched. It left a tangled mess of leveled houses, snapped trees, smashed boats, crumpled buildings, and mangled cars.

"The debris trucks ran seven days a week for seven months after Andrew," said one Homestead city council member.

North Americans call them hurricanes. In China and Japan they are known as typhoons; in countries around the Indian Ocean, they're called cyclones. To Filipinos and other Southeast Asians, they are baguios. No matter what they are called, these enormous spiraling weather systems are some of the greatest storms on earth. The United Nations' World Meteorological Organization estimates that in an average year, 80 of them kill up to 15,000 people worldwide and cause several billion dollars worth of property damage.

Hurricane Inez whips around the palm trees in Miami, Florida. *Getty Images*

Most North American hurricanes develop over the Atlantic within 20° of the equator. Each year hundreds of weather disturbances, or cyclonic storms, form in the easterly trade winds over the tropical ocean, but fewer than 10 become full-fledged hurricanes.

Although hurricanes are not fully understood, many factors are involved in the creation of these mammoth storms. In most cases, intense sunlight heats the ocean, which in turn warms the air by convection. The heated air rises, carrying away evaporated water charged with energy, and producing an area of low pressure. The process quickens as more and more air and water spiral upward. The air cools, condenses, and releases the sun's energy stored in the evaporated water. So awesome is the rate of heat-energy release inside a mature hurricane that if it could be changed into electricity, it would amount to 10 trillion kilowatts of power. That's about five million times the amount produced in the US on a given day.

It takes a complex series of chain reactions and self-feeding processes to create a hurricane. While the storm intensifies, the pressure in its central column is further lowered, creating a partial vacuum. More ocean air is sucked in and sent higher into the atmosphere as the whirlpool-like vortex spins faster. Within a few days, the minor disturbance that began innocuously on a calm, humid day is transformed into an enormous, whirling weather machine.

An important feature of the hurricane is the inward spiraling motion initiated by the Earth's rotation—a counterclockwise motion in the Northern Hemisphere and a clockwise motion

**Hurricane Mitch pounded Central America on October 29-31, 1998, and in just days, erased decades of progress from the maps of Honduras and Nicaragua, and to a lesser extent, El Salvador and Guatemala. In preliminary estimates, at least 11,000 lives were lost, and at least as many people were missing. It was the deadliest Atlantic hurricane in more than 200 years. Up to 75 inches of rain fell in places, causing deadly flash floods and torrential mudslides. By the time Mitch hit Florida on November 5, it had been downgraded to a tropical storm.**

## Name That Storm

Hurricanes and tropical storms are typically the only weather events with names, though in 1997 some Midwesterners started naming their blizzards.

Originally, hurricanes were sometimes named for the saint's day on which they occurred or for large ships sunk by the storm. At the turn of the century, Australian meteorologist Clement Wragge assigned female names to tropical storms, and male names, especially politicians', to other storms. However, for most of the first half of the century, hurricanes went unnamed.

The practice was revived during World War II when American Air Force and Navy meteorologists attached the names of their wives or girlfriends to hurricanes, especially when two or more storms appeared on the weather map at the same time. A system of latitude-longitude identification was also used but was considered too cumbersome for practical use, especially when more than one hurricane was brewing over the ocean.

During the 1940s, the media got into the hurricane game as well, calling the first storm in 1949 Hurricane Harry (Truman) and a subsequent, more violent, storm Hurricane Bess, after the president's wife.

in the Southern Hemisphere. Seen from above, the hurricane appears as a spiraling mass of cloud converging into a small area free of cloud. Outside this central column, the strongest winds can roar at more than 100 mph, accompanied by a constant deluge.

**The deadliest hurricane in history struck the Bay of Bengal in 1737; 300,000 people drowned. In 1991, more than 125,000 people were killed in a cyclone in Bangladesh. The disaster left 10 million people homeless.**

Paradoxically, at the heart of a storm exists a world of calm and sunshine in what is called the "eye." Many people have been lured from shelter when the eye passed overhead, only to be caught by the violent winds from the opposite side of the hurricane.

Other identification systems survived a year or so, for example, naming hurricanes by letters of the alphabet, as in the radio code words Able, Baker Charlie, and so on. One system even used a new international phonetic alphabet.

In 1953, American military communicators suggested using female names in alphabetical order. The practice was adopted by the weather service, despite thousands of letters of complaint. Weather sexism continued for 24 years until 1978. Although the weather service contended that mail favoring the feminine name system far exceeded that against it, in 1979 the World Meteorological Organization initiated the use of a pre-selected list of female and male names organized alphabetically. There are separate lists for Pacific (one each for eastern, central, and western Pacific cyclones) and Atlantic hurricanes.

The practice is still in use today. Short, distinctive English, French, and Spanish names are used and repeated every six years.

Names of devastating hurricanes are not recycled; they are replaced with new names. So, there will never again be another Hurricane Andrew. Many people, no doubt, are thankful for that.

The main system of cloud, rain, and wind extends about 35 to 55 miles out from the center of the eye. Farther out, stretching over tens of thousands of square miles in spiral bands, conditions are less extreme, but winds can still be violent and rains heavy. At some point, the young hurricane starts to move, usually advancing at a speed of 10 to 15 miles per hour. While moving out of the tropics, it gradually accelerates, often wreaking havoc along the way.

But once the storm moves out of the tropical zone and into cooler northern latitudes, it quickly dies out—either starved of heat energy and moisture when it encounters cold seas or dragged apart by ground friction when it meets land. The farther north from the equator the storm moves, the more likely it is to get caught in the prevailing westerlies. Eventually, it curves again to the north and northeast.

## The 10 Deadliest Hurricanes

| | Location | Year | Category | # of Deaths |
|---|---|---|---|---|
| 1 | Galveston, Texas | 1900 | 4 | 8,000–10,000 |
| 2 | Lake Okeechobee, Florida | 1928 | 4 | 1,836 |
| 3 | Florida Keys & South Texas | 1919 | 4 | 600 |
| 4 | New England | 1938 | 3 | 600 |
| 5 | Florida Keys | 1935 | 5 | 408 |
| 6 | Louisiana & Texas | 1957 | 4 | 390 |
| 7 | Northeastern US | 1944 | 3 | 390 |
| 8 | Grand Isle, Louisiana | 1909 | 4 | 350 |
| 9 | New Orleans, Louisiana | 1915 | 4 | 275 |
| 10 | Galveston, Texas | 1915 | 4 | 275 |

Source: National Hurricane Center

That is why the most hurricane-prone regions in the US are the southeastern coast—Florida and the Carolinas—and the Gulf coast states—Texas, Mississippi, and Louisiana.

The average lifetime of a hurricane is nine days. Andrew was born off the coast of Africa on August 17, 1992. It grew to hurricane strength on August 22, and was seven days old when it ravaged southern Florida. Some hurricanes, however, may last only a few hours, while others continue for almost a month.

The North Atlantic hurricane season extends from June through

## The 10 Costliest Hurricanes

| | Location | Year | Category | Damage |
|---|---|---|---|---|
| 1 | Andrew, Florida & Louisiana | 1992 | 4 | $30,475,000,000 |
| 2 | Hugo, South Carolina | 1989 | 4 | $8,491,561,181 |
| 3 | Agnes, Northeastern US | 1972 | 1 | $7,500,000,000 |
| 4 | Betsy, Florida & Louisiana | 1965 | 3 | $7,425,340,909 |
| 5 | Camille, Mississippi & Alabama | 1969 | 5 | $6,096,287,313 |
| 6 | Diane, Northeastern US | 1955 | 1 | $4,830,580,808 |
| 7 | Frederic, Alabama & Mississippi | 1979 | 3 | $4,328,968,903 |
| 8 | New England | 1938 | 3 | $4,140,000,000 |
| 9 | Fran, North Carolina | 1996 | 3 | $3,200,000,000 |
| 10 | Opal, Florida & Alabama | 1995 | 3 | $3,069,395,018 |

Source: National Hurricane Center

November, although 84 percent of the storms occur between August and October. They reach their greatest fury and frequency in September, when the surface ocean temperature is at a peak of around 80°F.

Of the 10 tropical cyclones that develop each year in the north Atlantic, only six will become hurricanes. A tropical storm becomes a hurricane when its wind speed reaches 74 miles per hour.

**Hurricane Allison holds the record as the deadliest and most expensive tropical storm in US history. It developed on June 4, 2001, and drifted back and forth over land and sea for more than two weeks. Allison caused extremely heavy rainfall, resulting in severe flooding all the way from eastern Texas to the mid-Atlantic coast. Forty-one deaths throughout six states were reported. Damage estimates after the storm reached more than $5 billion.**

The most destructive elements of a hurricane are often the huge waves, storm surges, flooding, and landslides it generates. Sea levels are pushed up as much as five yards higher than normal tides, flooding low-lying coastal areas. Hurricane winds have produced waves as tall as 10-story buildings; a US navy ship's commander reported encountering a wave over 100 feet high in 1933. And while the direct effects of a hurricane can be devastating, the tornadoes they frequently spawn, which have even higher wind speeds, can also take a heavy toll.

Like tornadoes, the intensity of a hurricane is rated. The Saffir-Simpson scale ranks them from Category 1 to 5. Category 5 hurricanes have the highest sustained winds—over 155 mph—and the lowest barometric pressure. They inflict the most damage.

The two most powerful hurricanes to hit the US this century were Category 5: Hurricane Camille in 1969, and an unnamed storm over the Florida Keys on Labor Day 1935. Andrew was the third strongest, but was rated only a Category 4, though its winds were estimated to be the strongest for a landfall hurricane. Andrew may have even intensified as it struck land, which is not normal. Its winds gusted up to 175 mph, and possibly even 200 mph.

## The Saffir-Simpson Hurricane Scale

| Category | Wind speed | Storm surge (feet) | Barometric pressure (inches) | Typical damage |
|---|---|---|---|---|
| 1 | 74–95 | 4–5 | 28.94 or higher | **Minimal**: No real damage to buildings. Shrubs, trees, and unsecured mobile homes damaged |
| 2 | 96–110 | 6–8 | 28.50–28.91 | **Moderate**: Considerable damage to trees, shrubs, and mobile homes. Some damage to roofing material, doors, and windows |
| 3 | 111–130 | 9–12 | 27.91–28.47 | **Extensive**: Foliage torn from trees and shrubs. Some damage to small homes and buildings. Mobile homes destroyed |
| 4 | 131–155 | 13–18 | 27.17–27.88 | **Extreme**: Shrubs and trees blown down. Extensive damage to roofing material, doors, and windows. Complete roof failure on small homes. Mobile homes completely destroyed |
| 5 | >156 | >18 | <27.17 | **Catastrophic**: Considerable damage to roofs. Very severe window and door damage. Complete roof failure on many buildings. Some complete buildings fall |

Scientists can track hurricane development on Doppler radar, and they use observations from a high-altitude Gulfstream jet to model potential hurricane movements. Buoys in the ocean record surface temperatures, and satellites gather a wealth of atmospheric data. Though scientists can make forecasts, they cannot always predict precisely where the hurricane will go, or how strong it will get.

"Hurricanes don't become killer storms by slow and steady improvement," said Hugh Willoughby, of the Hurricane Research Division of the NOAA in Miami. "There's a fast track. They can go from a Category 2 to Category 5 in a day."

This makes predicting their potential devastation difficult, and knowing when to alert people even more difficult. False alarms are like crying wolf. A hurricane evacuation in a densely populated area is not only expensive; if it's a false alarm, forecasters and disaster management officers also lose credibility, and people might not listen to their warnings the next time a severe storm threatens. When asked how to know when to make the decision to evacuate, Hugh Willoughby replied: "You sweat."

Luckily, by the time Andrew hit south Florida, much of the area had been evacuated. Others had prepared for the wrath by hunkering down inside their homes. The death toll was 43, which is low for a storm of Andrew's intensity, and credit for the low loss of life goes to forecasters and disaster management officers.

But nothing could stop the property damage that Andrew inflicted. Its spiraling winds and powerful surges wrecked 80,000 homes. In terms of damage, it was the granddaddy of all hurricanes. It caused between $25 billion and $30 billion in damages—the costliest single disaster ever to strike the US.

The damage Andrew did showed the vulnerability of weaker structures, and drew

**The Labor Day Storm of '35: The most powerful Atlantic hurricane of this century smashed into the Florida Keys in the early hours of September 2, 1935, leaving over 400 people dead. Storm surges were the big killer. Some victims were actually sandblasted to death. More than half of the dead were World War I veterans who had been stranded in their meager quarters near the edge of the ocean. Ernest Hemingway, who had a house on Key West, was appalled by the death and destruction he saw following the storm. He believed the government was to blame for not moving the former soldiers before the storm, and wrote a condemning article called "Who Murdered the Veterans?"**

**Atlantic Tropical Storm and Hurricane Names**

| 1999 | 2000 | 2001 | 2002 | 2003 |
|---|---|---|---|---|
| Arlene | Alberto | Allison | Arthur | Ana |
| Bret | Beryl | Barry | Bertha | Bill |
| Cindy | Chris | Chantal | Cristobal | Claudette |
| Dennis | Debby | Dean | Dolly | Danny |
| Emily | Ernesto | Erin | Edouard | Erika |
| Floyd | Florence | Felix | Fay | Fabian |
| Gert | Gordon | Gabrielle | Gustav | Grace |
| Harvey | Helene | Humberto | Hanna | Henri |
| Irene | Isaac | Iris | Isidore | Isabel |
| Jose | Joyce | Jerry | Josephine | Juan |
| Katrina | Keith | Karen | Kyle | Kate |
| Lenny | Leslie | Lorenzo | Lili | Larry |
| Maria | Michael | Michelle | Marco | Mindy |
| Nate | Nadine | Noel | Nana | Nicholas |
| Ophelia | Oscar | Olga | Omar | Odette |
| Philippe | Patty | Pablo | Paloma | Peter |
| Rita | Rafael | Rebekah | Rene | Rose |
| Stan | Sandy | Sebastien | Sally | Sam |
| Tammy | Tony | Tanya | Teddy | Teresa |
| Vince | Valerie | Van | Vicky | Victor |

attention to lax building standards in southern Florida. Homestead had 1,200 mobile homes before Andrew. After Andrew, just eight were left standing.

Nature had a crushing effect on itself: Andrew decimated trees and forests. It wiped out some 90 percent of the pine trees in Dade County. The trees either died from being blown over or broken by the storm or from a beetle infestation that followed.

"There's not a tree out there tall enough for a self-respecting eagle to land in," said a Homestead town council member.

Andrew killed people and pets and plants and livestock, and in its path of destruction, it left 160,000 people in southern Florida homeless. In the weeks after it passed, several thousand residents would leave the area and never return. Those left behind would have to rebuild their shattered lives, and prepare themselves and their property for the next hurricane.

And Homestead? Its population is growing again. Businesses have been re-opened, homes rebuilt. It hasn't fully recovered from Andrew, though, and it has lost years of growth and development.

"We never thought it would take this long to recover," a town council member said on a warm December day, more than five years after Andrew obliterated her community. "We've still got a couple of Beirut neighborhoods."

**The first "male" hurricane in the Atlantic was actually a wimp. In 1979, Hurricane Bob barely made it over the 74 mph speed limit to qualify as a hurricane. In 1991, though, Hurricane Bob was deadly and the name was retired.**

# Tornado's Fury

On a dark day in April 1899, two women and a young boy were plucked into the air by a mysterious force. They flew about a quarter of a mile. They remembered seeing a church steeple passing below. Most amazingly, they were accompanied in their flight by a horse, which kicked as it flew. The horse flew about a mile farther than they did, and was also put down gently.

Was this some form of alien abduction? No. It was a manifestation of the weirdest and most dangerous threat in weather's arsenal: the tornado.

It can happen in any state, at any time. It might swirl for a benign minute in the middle of an empty field, or it might twist savagely through a busy town, picking up cars, horses, and houses and leaving nothing but shards in its deadly wake. It can happen on the plains or in the mountains, in cities or forests, early in the morning or late in the evening. It's a tornado, and it's predictably unpredictable.

In the past century, more than 10,000 Americans have died in tornadoes. About 1,000 tornadoes are recorded each year in the US—over 10 times more than in any other country. Since 1950, on average, 88 people have died each year in US tornadoes. This number is declining, though the death toll in 1998, which was 129, was much higher than normal.

## Mr. Tornado

- Dr. Tetsuya Fujita, who passed away on November 19, 1998, was nicknamed "Mr. Tornado." Before his death, Dr. Fujita had studied every recorded tornado in the 20th century in the US. He devised the Fujita classification for tornadoes. He also discovered microbursts—intense downdrafts that can spawn winds as strong as 150 mph on the ground. The destruction he studied in the 1973 Super Tornado Outbreak reminded him of what he had seen at Hiroshima and Nagasaki in 1945.
- The US gets far more tornadoes per year than any other country—about 1,000. Canada ranks second, with over 80 per year, followed by Russia with 60, Great Britain with 40, Australia with 15, and Italy with 10.

Tornadoes aren't like hurricanes, which brew over open waters and can take days to reach land. They are born from thunderstorms, which form when warm, humid air meets a mass of cool, dry air. Only one in 100 thunderstorms produces a tornado. Tornadoes happen so quickly, and are so violent, that scientists can't fly into them, as they can with hurricanes. Much of the research, therefore, is often done in laboratory simulations, at places like the National Severe Storms Lab and the University of Oklahoma.

In 1994–95, Dr. Erik Rasmussen, a prominent tornado researcher, led a team of more than 100 tornado researchers in a project called VORTEX (Verification of the Origins of Rotation in Tornadoes Experiment). VORTEX became the largest tornado observation program ever. Researchers followed storms in Texas, Oklahoma, and Kansas and used portable Doppler radar to collect information on the wind speed and direction of tornadoes. They videotaped and recorded atmospheric conditions. Data gathered during VORTEX are being used to test hypotheses and to develop new theories of how tornadoes are formed and why they twist.

Though scientists aren't sure of the precise way in which tornadoes are formed, the basic conditions by which tornadoes form are understood. Inside a thunderstorm, warm air rises. As the warm, moist air meets cool air, the moisture condenses. The process of condensation itself releases energy, called "latent heat," which further warms the air. The heated air then rises faster, breaking through the layer of cold air, which creates instability and turbulence inside the thunderstorm.

There are literally a dozen hypotheses on how a tornado is formed inside a thunderstorm, including one that suggests that a tornado may be born in a swirl of air outside the storm, then amplified by the storm's power. Some scientists think that when the warm updraft from the thunderstorm's core meets the cool downdraft, a spinning column of air—called a vortex—is formed, almost the way a hand sweeping across a blanket can roll up a twist of lint. This vortex is eventually tilted down vertically, perhaps by the downdraft. It's possible that the downdraft is what causes the tornado to spin. Some have theorized that a spinning vortex is "stretched" into a tornado, either by an updraft, or by the meeting of a downdraft and updraft. Others think that a vortex works like a vacuum, and sucks up air from below, until the column is long enough to reach the ground.

Regardless of what causes the vortex to spin, or what causes the rotating column to become vertical and reach the ground, the spinning air gathers speed as the tornado tightens. Many compare this process to the way a figure skater spins faster as she tucks her arms close to her body. As the speed increases, so does the tornado's ferocity.

Some tornadoes spin as slowly as 40 mph, but others might reach speeds of up to 300 mph. The average forward speed of the rotating columns is 35 mph, but they can race ahead at more than 60 mph. A 1930 tornado in Kansas strolled along at just 5 mph, but that's extraordinarily slow. Nobody advises trying to outrun a tornado, because they can pick up speed very quickly. Some people try to drive away from a tornado, and sometimes this saves their lives, but in general, a car is a dangerous place to be when a tornado strikes.

Tornadoes generally move from the southwest to the northeast, but they can move in any direction, and have even been known to travel in circles or make U-turns. A 1944 tornado traveling northwest in Iowa paused and twisted on one spot for several minutes, then went southeast, then south, then east, then north, and then east again.

Tornadoes usually stay on the ground for only minutes, but sometimes they grind up landscape for much longer. In 1917, a tornado that started in Missouri stayed on the ground for seven hours and 20 minutes and cut a path 293 miles long from its starting point on through Illinois and Indiana.

As a tornado moves, it gathers dust and debris, which gives it its color. “It looked like a big black mushroom,” said a Texas woman who watched as a tornado raced toward her house. A tornado picks up objects, hurling and twirling them around in a vortex. Powerful tornadoes can carry heavy things hundreds of yards. In one tornado a motel sign was carried 30 miles from Oklahoma to Arkansas.

In strong tornadoes, houses can be pulverized, lumber splintered, railcars thrown off the tracks, and cars twisted into unrecognizable hunks of metal. During a 1995 tornado in Pampa, Texas, there were reports of six vehicles twisting through the air at the same time. In April 1963, a Mississippi tornado tore pavement up from a highway and carried it hundreds of yards. In March 1998, a tornado in Hall County, Georgia, lifted a house from its foundation and carried it 200 feet before dropping it. It splintered when it fell, but, amazingly, a man, his wife, and two young children inside survived.

Tornadoes are classified according to their wind speed and the damage they inflict. The late Dr. Ted Fujita, who researched tornadoes at the University of Chicago, devised a system for measuring the force of tornadoes. He decided to calculate a tornado's intensity by examining the damage and debris after the tornado hit. Depending on the type of damage, a tornado is ranked from F0 to F5, with F5 being the most severe.

Under the Fujita classification, about three-quarters of tornadoes are weak (F0–F1), about one-quarter are strong (F2–F3), and about one to two percent are classified as violent (F4–F5). Not surprisingly, the strongest tornadoes cause the most death and destruction; F4 and F5 tornadoes account for over two-thirds of tornado deaths. In 1998, there were two F5 tornadoes, one in Waynesboro, Tennessee, which took three lives, and one in April in western Alabama, in which 34 people died, 256 were hurt, and over a thousand homes were destroyed or damaged.

A strip of land that extends from northeastern Texas through

## The Fujita Scale

| Category | Description | Wind speed | Examples of observed damage |
|---|---|---|---|
| F0 | Light | 40–72 mph | Some damage to chimneys, breaks twigs and branches off trees, pushes over shallow-rooted trees, damages signboards, breaks some windows |
| F1 | Moderate | 73–112 mph | Peels surfaces of roofs, pushes mobile homes off foundations, pushes moving cars off road, snaps some trees, demolishes outbuildings |
| F2 | Considerable | 113–157 mph | Tears roofs off frame houses, demolishes mobile homes, some frame houses lifted and moved, large trees snapped or uprooted, light objects become missiles |
| F3 | Severe | 158–206 mph | Most trees uprooted, cars lifted and thrown, roofs and some walls torn off well-built houses, trains flipped over, some pavement ripped off roads |
| F4 | Devastating | 207–260 mph | Well-built houses leveled, uprooted trees carried away, buildings with weak foundations blown off, cars thrown and disintegrated |
| F5 | Incredible | 261–318 mph | Strong frame houses lifted and carried away and disintegrated, car-size missiles fly more than 300 feet, trees debarked |

Oklahoma, Kansas, Nebraska, and Missouri has more tornadoes than anywhere else in the US. It is known as "Tornado Alley." From 1950–1995, Texas got an average of 124 each year. Over the same time period, Rhode Island averaged less than one tornado per year. Although the incidence of tornadoes in Tornado Alley is highest in the country, more people have died in tornadoes in the Southeast over the past 50 years than in any other region.

By state, Florida gets the most tornadoes per square mile. In 1998, Florida suffered its deadliest tornado tragedy: 42 people were killed and 260 hurt in a late-night outbreak on February 22–23. The American Southeast was especially hard hit in 1998: the region accounted for 104

## You've Come a Long Way . . .

- A bond deed was carried 125 miles through the air in the terrible Tri-State Tornado in Missouri, Illinois, and Indiana in 1925. The finder mailed it back to the owner, a policeman from Murphysboro, Illinois.
- Canceled checks flew over 200 miles in a tornado from Texas to Oklahoma in 1979.
- Following a deadly tornado in New Baltimore, Michigan, at the end of April 1964, debris began showing up a few days later in Canada in the Strathroy area of Ontario. Among the found articles were canceled checks from the Citizens State Savings Bank, retail sales receipts from a lumber company, a court summons, and a boy's sport jacket.
- A postcard was carried 122 miles from Indiana to Ohio in a 1922 tornado. It was not, however, delivered to the right address.
- A Saskatchewan gentleman lost his trousers in a June 1923 twister, and later, to his surprise, found them swinging from a tree over a mile away.
- Even the cities aren't safe. A common misconception is that tornadoes don't hit big cities. But that's not true: Chicago, Miami, St. Louis, Austin, Houston, Minneapolis-St. Paul, Cincinnati, Louisville, Nashville, Indianapolis, Omaha, and Raleigh have all been hit by tornadoes. Dr. Fujita's theory was that some small tornadoes might be suppressed by the "heat island" from a city, but scientists report that this theory would not hold true for strong tornadoes.

of 129 US tornado casualties. In spite of the bad reputation and past experience of Texas, only two people were killed by tornadoes there in 1998. Oddly, the state that recorded the most tornadoes in 1998 was Illinois, with 104; Texas had just half as many, 52.

Tornadoes have touched down in every state, on all types of terrain. A tornado stormed through Yellowstone National Park in the late 1980s and tore away part of a 10,000-foot mountain. In 1944, four simultaneous tornadoes killed 153 people in mountainous regions of Pennsylvania, West Virginia, and Maryland. Tornadoes occasionally touch down in California—in 1988, a tornado picked up a baseball dugout in Orange County and carried it 150 yards. A tornado was even reported above the Arctic Circle, at Kiana, Alaska.

Tornadoes don't always happen in isolation. Often multiple tornadoes grow from a single thunderstorm. The most tornadoes to appear in one squall line were 148, on April 3–4, 1974. This terrible swarm, known as "The Great Outbreak," included six F5 tornadoes and terrorized people from Alabama to Michigan and even into Canada. Three hundred and fifteen people died, and over 5,000 were hurt.

These kinds of multiple tornadoes are not unusual. In September 1967, Hurricane Beulah dragged 115 tornadoes in her wake as she roared through Texas. The 42 deaths in Florida on February 22–23, 1998, were caused by a series of seven tornadoes ranging from F2 to F3 in intensity. On October 4, 1998, 20 tornadoes touched down in Oklahoma, setting a record for the most twisters on an October day.

The lifting power of a tornado is incredible. A 13-ton tank was moved three-quarters of a mile by a deadly tornado in Lubbock, Texas, in 1970. Seven train cars were plucked from their tracks in a 1919 tornado; the baggage car landed 30 feet from the others. A 500-pound baby grand piano sailed several hundred feet through the air in a 1973 Nebraska tornado. A church steeple flew 15 miles in a tornado.

Tornadoes are powerful—and destructive—for the same reason an airplane flies. Wind moving over a curved object picks up speed. As it moves faster and faster over the surface of the object—in the plane's case, the wing—it creates an area of low pressure above the wing, which creates lift. The strong wind in a tornado can do the same with houses, cars, and trains. The tornado's fierce wind literally lifts the roof off and, depending on the intensity of the wind, makes deadly missiles out of objects as small as shingles or as large as cars.

The danger of being in a mobile home during a tornado is legendary. The weaker the foundation, the easier it is for a tornado to lift a home, which is one reason mobile homes are so vulnerable. Over the past 12 years, 38 percent of the people killed in US tornadoes have been in mobile homes.

### Repeat Offenders

- There's a southwest-to-northeast path from St. Louis, Missouri, across the Mississippi River to Granite City, Illinois, that seems to be especially prone to tornadoes. Deadly tornadoes have taken that route in 1871, 1896, 1959, and 1927.
- Tornadoes hit Codell, Kansas, on May 20, 1916, again on May 20, 1917, and again on May 20, 1918.
- Thirty-one tornadoes ravaged the South on March 21, 1932. Twenty years later, on the same date, 31 tornadoes hit the South again.
- May 11 is a sad day for Texans: on that date in 1953, a tornado hit Waco, killing 114 people. On the same date in 1970, a powerful tornado hit Lubbock. Twenty-eight people died and half the city was wrecked.
- One church in Arkansas has been hit by tornadoes three times this century.

Disaster management officials advise people to get out of mobile homes when there's a tornado warning, and find shelter underground or in the basement of a sturdier home. However, in an unpredictable tornado, anything's possible. In the 1997 Jarrell, Texas, tornado, two teenage brothers biked from their modular home to the apparent safety of a neighbor's sturdier frame house. The frame house was destroyed, and both boys died, along with four members of the neighbor's family. The modular home they had fled was not destroyed.

Unfortunately, in a tornado as powerful as the F5 in Jarrell—which was especially damaging because it twisted fiercely but moved very slowly—sometimes there simply aren't any safe places. "It just sat there and chewed and chewed," said a researcher from Texas Technical University in Lubbock.

The combination of rapidly rotating wind and flying debris can make a tornado deafening.

"It sounded like 100 trains," said a woman who lived through the 1997 Jarrell tornado. "It was like saws in a rock quarry," said another.

"It was like a bomb," survivors of the Jefferson County, Alabama, tornado in April 1998 told newspaper reporters. In Manila, Arkansas, on April 16, 1998, sirens were sounded to alert the people of a powerful F4 tornado, but the sounds of thunder and rain and wind were too great. Two people died. But sometimes the noise of a tornado is a blessing: in 1959, a powerful tornado that moved slowly through Oklahoma and Kansas was so loud it alerted people it was coming. Nobody was killed.

Regardless of the intensity of the tornado, there are strange tales of what they destroy and what they spare. They are overwhelmingly fickle.

The F5 tornado in Jarrell on May 27, 1997, was the strongest and deadliest of the year. Twenty-seven people died, and 40 homes were completely destroyed—some carried away in shreds, leaving bare concrete slabs behind. One woman crouched in the bathtub with her family as the tornado tore apart their house. They survived "without a scratch."

After the tornado was over, and after she and her family had pried themselves from the mud and debris, they went outside.

"There wasn't anything left," she said later. "Nothing. The houses were gone. No vehicles, no refrigerator, no washing machine. Where did it all go?" Then she found her stack of photo albums, which she had stored in her linen closet. They were perfectly intact and stacked exactly the same way she'd left them. They were 20 feet away from what remained of her house.

Another family, who lived about a mile away, reported that their pickup truck was carried 75 feet. "The tornado mulched everything," one member of the family said. Before the tornado, they had stacks of 2×4 lumber in their yard; afterwards, there were just splinters. But a reclining chair was sitting upright in the middle of their field, intact.

In nearby Cedar Park, which was also hit by the incredible tornado, neighbors found a bible with a marriage license inside, which belonged to a couple whose house was destroyed a block away.

A February 1998 tornado in Kissimmee, Florida, hurled an 18-month-old baby lying on his mattress from his great-grandmother's house into the branches of a fallen oak tree. Incredibly, the baby survived, still swaddled in his mattress. Another lucky baby suffered just minor injuries after being carried some 300 yards in a Louisiana tornado in 1973. On March 8, 1998, a newlywed couple was blasted from a closet in their Alabama home; the tornado carried them half a block from their demolished house, and both survived.

Since the first tornado forecast over 50 years ago, much research has led to better technology and forecasting methods. Doppler radar has made tracking tornadoes easier, but since tornadoes develop so quickly and move so erratically, there isn't always enough time for everyone to prepare for their destructive wrath. And the technology doesn't always catch tornadoes: 12 people in Georgia died in a March 1998 tornado that wasn't detected by Doppler radar.

However, increased public awareness and willingness to prepare for disaster has, in general, been accompanied by fewer tornado deaths. The 25 deadliest tornadoes in the US all happened in 1955 or earlier. In the 1930s, an average of 195 people died each year in US tornadoes; by the 1980s, an average of just 52 people died each year in tornadoes.

Scientists and disaster management officials are able to make generalizations about tornadoes, which helps people to understand them better, or at least to prepare for them more effectively.

For instance, it's known that about 80 percent of tornadoes strike between noon and midnight. The conditions under which tornadoes are formed are known, and Doppler radar, which allows scientists to monitor atmospheric movements, can potentially identify air rotations that might result in tornadoes. The development of the advanced radar system, NEXRAD, allows scientists with the National Weather Service and the Department of Defense to observe the atmosphere up to 143 miles away from the position of the radar. Sophisticated technology like this, combined with efficient communications like radio, TV, and telephone, can give people time to go to shelters and other safe places.

This is not just theoretical. This new knowledge has already saved lives. In Jefferson County, Alabama, on April 8, 1998, a couple in a mobile home heard a tornado warning on TV, so they ran to their neighbor's frame home. They survived, while their mobile home was totally wrecked.

However, one of the difficulties in trying to warn people is that tornadoes form so quickly. A National Severe Storms Laboratory meteorologist estimates that tornadoes form within five to 10 minutes, so

## Killer Twisters

- The deadliest US tornado pounded Missouri, Illinois, and Indiana on March 18, 1925. The "Tri-State Tornado" traveled an incredible 219 miles. By the time the F5 tornado was finished with its cruel itinerary, 689 people lay dead, 2,000 were hurt, and thousands were homeless. The Illinois town of Murphysboro, where 234 people died, was hardest hit. There were, however, some happy stories: 16 students were swept up and carried more than one hundred yards, but were set down unhurt.
- This century's second deadliest tornado, also an F5, ripped through Tupelo, Mississippi, on April 5, 1936. Two hundred and sixteen people died, along with countless chickens. One of the young survivors of that tornado grew up to be famous: Elvis Presley. Many years later, on August 29, 1978—after The King had died—a tornado missed Graceland by less than a mile.

depending on how far centers of population are from the tornado, there may be very little time in which to notify people.

And the thunderstorms that spawn tornadoes can themselves cut power and communications systems. On April 8, 1998, a St. Clair, Alabama, couple lost power and so had no TV, radio, or phone. No warnings of the approaching tornado reached them, and they were killed. In Spencer, South Dakota, on May 30, 1998, a loss of power apparently caused the outdoor fire siren to fail. Six elderly Spencer residents died in the F4 tornado.

Tornado research, public education, and improved communication have led to reduced death tolls, but the unpredictable nature of tornadoes—and the brutal and unbelievable strength with which they can ravage a community—leave people, animals, and property vulnerable. But though tornadoes are merciless as they indiscriminately kill, hurt, demolish, and wreck, people remain fascinated by them.

"It was amazing," said a woman in Jarrell, six months after the bloodthirsty F5 tornado roared through her town and took the lives of 27 members of her community. "It was amazing. I never want to go through it again, but it was an amazing experience."

## Chasing the Fury

There are actually people who seek out tornadoes; they're called storm chasers.

Storm chasing is not a new pursuit. It began among meteorologists using airplanes in the United States in 1948. Ground-based chasing took hold in the 1950s, when the network of paved roads expanded, enabling storm chasers to move faster than most storms, which typically travel at speeds below the posted limit. One of the first successful ground-based chases occurred on May 24, 1973, when scientists from the US National Weather Service and the University of Oklahoma intercepted and issued warnings of a tornado west of Oklahoma City. For the first time, meteorologists were able to record the entire life cycle of a tornado on film and Doppler radar.

In recent years, chasing storms has developed into a craze, crowding the highways on weekends with amateur spotters and television news teams, some in helicopters. Armed with video cameras, thrill seekers and adrenaline junkies prowl the central United States hoping to capture spectacular weather footage to sell to network television, seeking fame and fortune by selling videotapes and photographs.

Tornado Alley is the ideal location to watch for storms. It's big-sky country—clear and open—without the haze and pollution of industrial regions, and away from the obstruction of mountains.

"Everyone asks why do I do it," says one storm chaser. "I know some of my relatives think I'm foolish, and there are meteorologists who don't understand why chasing storms can be so compelling."

All serious storm chasers have a healthy respect and abiding curiosity for summer storms. It's not just thrill seeking. Storm chasers have contributed greatly to our understanding of the structure and movement of summer storms and the conditions likely to produce them. Photographing and observing summer storms has also been useful for training severe-weather watchers and familiarizing the public with the appearance of extreme summer storms.

The June 1991 eruption of Mount Pinatubo in the Philippines caused temperatures to fall around the world. *Ponopresse*

Snowrollers are giant natural snowballs formed by the wind. *R. S. Schemenauer*

Sun pillars are most common at sunset and sunrise. *R. S. Schemenauer*

The Blizzard of '96 blinds pedestrians on Seventh Avenue in New York. *Canapress*

Ice Storm '98 left millions of people without electricity, some for weeks. *Paul Norbo*

A perfect rainbow stretches across the Prairie sky. *First Light*

Tens of thousands of people were forced to leave their homes during the Red River flood of 1997. *Ponopress*

A tornado touches down near the small town of Saint Stanislaus. *Stephen Dupont/Canapress*

Storm chasing involves much more than hopping in a car and watching the sky. The storm coordinator alerts other members to the possibility of severe weather the day before, and preparations for the chase get under way immediately. The storm coordinator forms one or two chase teams that ideally comprise three people—a driver, a navigator, and a photographer. Each team then lines up a vehicle, buys film, charges the batteries, and puts together a storm kit that includes a cellular telephone, still and video cameras, maps, a compass, log sheets, and portable weather equipment provided by the weather office—an anemometer to measure wind speed, a rain gauge, and thermometers.

The next day, the team heads to the office to check the morning's weather charts, radar scans, and satellite images. Meanwhile, the storm coordinator and severe-weather meteorologist monitor the hourly weather changes and pore over the upper-air soundings. A successful chase depends on the coordinator's correctly pinpointing the chase field—the area where a thunderstorm is likely to occur. Then the

## You Lucky Dog . . .

- On July 9, 1987, a tornado in Michigan churned through a dog-boarding kennel. Later, one of the dogs was found high in a tree half a mile away—unhappy, no doubt, but unharmed.
- Another pooch was lifted, in his doghouse, in an F2 tornado in Iowa in 1994. The dog was carried in his house several blocks, then set upside-down. He survived.
- In Westlock, Alberta, on June 5, 1956, a bull weighing some 1,400 pounds was resting in his stable, attached to the stall by a logging chain. After the tornado passed, the stall was found 60 feet away. The bull, apparently unhurt, was still attached.
- A July 1909 tornado in Saskatchewan tore all the shingles off one side of a school, raced across a road, took the roof off a barn, crossed the road again, and lifted a horse about 30 feet before setting him down unharmed.
- Five horses were picked up in a 1915 Kansas tornado that obliterated their farm and killed two people. The fortunate horses were carried a few hundred feet and set down, unharmed and still tied to a rail.
- A cow in Saskatchewan was found lying on her back, four legs in the air, anchored to the ground by her horns after a tornado struck in July 1946. The cow was otherwise unhurt, but the farmer chose not to milk her because he thought the milk would be sour.
- Following a deadly Woodstock, Ontario, tornado on August 7, 1979, a live pig was found stuck in a fork of a tree, where the wind had carried it.
- A woman was vacuumed out of a truck in a Midwestern tornado in 1990. During the tornado she saw a live deer fly past her.
- In September 1894, a 256-square-foot hen-house was carried away by a tornado and jammed between two trees. The diligent hens were still sitting on their intact eggs when they were found the next day.

### And the Unlucky Ducks . . .

- Tens of thousands of ducks died at a bird refuge in a Kansas tornado in 1915. The dead fowl fell from the sky many miles away. In the same tornado, 1,000 sheep met their untimely deaths.
- A Wisconsin tornado killed 10,000 turkeys in 1994.
- An alligator fell from the sky during a storm in Charleston, South Carolina, in 1843.
- Near Belleville, Ontario, a severe thunderstorm and tornado spread millions of tiny toads, each about the size of "a three-cent piece," across the ground for several miles in June 1870. Local residents said that the same phenomenon occurred on four other occasions.
- A tornado in Hall County, Georgia, on March 20, 1998, destroyed several large commercial chicken houses. Up to 100,000 chickens died, while other lucky ones, liberated by the tornado, flew the coop.

coordinator must decide if the storm is accessible and within safe driving distance. He or she arrives at a decision, usually by mid-morning, whether to dispatch teams to a designated target area, to put them on standby, or simply to call them off.

Once a team gets the go-ahead, the navigator chooses the intercept route on the basis of road availability, visual storm observations, and information received from the weather office. The navigator is also responsible for documenting the chase—landmarks, phone calls, starts and stops, and meteorological sightings. The driver's responsibility is to operate the vehicle in a safe and legal fashion—and to pay any traffic tickets. Readying the vehicle for a quick exit is also important.

Storm chasing is largely a driving and waiting game. Inside the car, the team members constantly read the sky for signs of atmospheric change, take weather measurements, keep in touch with the weather office by cellular phone, and listen for static on local radio stations that may signal an approaching storm. The talk is often of tail-end Charlies and rear-flank downdrafts, outflows and gustnadoes, and anvils and

## The Most Expensive Tornadoes

| | Date | Place | Property damage (1995 $US) |
|---|---|---|---|
| 1 | May 6, 1975 | Omaha, Nebraska | $1.132 billion |
| 2 | April 10, 1979 | Wichita Falls, Texas | $840 million |
| 3 | May 11, 1970 | Lubbock, Texas | $530 million |
| 4 | June 8, 1966 | Topeka, Kansas | $470 million |
| 5 | October 3, 1979 | Windsor Locks, Connecticut | $420 million |

shears. In many cases, chasers spend hours of frustration and disappointment parked at the side of a country road, often in sultry air and under blue skies, waiting for something to happen that probably won't.

Robert Davies-Jones, director of the tornado-intercept project at the United States National Severe Storms Laboratory in Norman, Oklahoma, estimates that he spends about 200 hours in the field each storm season.

"We hope for one or two good days," he says, "and in that, two to three good hours. And then it boils down to about 10 minutes, as far as a tornado is concerned. And that's the whole season."

Chasers try to convince one another of the good signs—ragged cloud peaks, sharp anvils, emerald green skies, clusters of cumulus towers that appear to grow and die, multiple layers of low and middle clouds racing overhead in different directions, and a few drops of rain.

Despite the long delays and disappointments, storm chasers can never allow frustration to overcome caution and good sense. Storm chasing can be both difficult and dangerous, and accidents do happen.

Storm chasers consider lightning and hail—not tornadoes—the greatest hazards associated with their hobby. Setting up an aluminum tripod on a hilltop is not a smart thing to do with lightning flashing all around you. Careful chasers minimize the risks by taking photographs from inside their car at the first drops of rain. In 1981, a lightning bolt struck three storm chasers in central Oklahoma. All three survived, but they had headaches for hours afterward. Car accidents are another hazard. In 1984, a graduate student from the University of Oklahoma was killed when his car overturned on a rain-slicked highway while he was pursuing a tornado. As a rule, chasers usually avoid unpaved roads. Dirt roads are just too treacherous when wet.

Other chase rules include these: don't hunt alone; if you can't see the cloud and ground, you are too close; where there is one tornado, there is possibly a second one nearby; and never drive blindly into the precipitation core—there could be a tornado right behind it.

By persevering in the face of these risks, trained storm chasers have contributed most of what we know about the size, speed, and internal structure of tornadoes and the conditions likely to produce them. They were the first to show that although most tornadoes rotate counterclockwise, some spin clockwise and develop in a different part of the storm system. This has in turn improved forecasting, because meteorologists now know to look elsewhere for tornadoes. Storm chasers were also the first to determine that tornadoes typically develop in a wall cloud—a low-hanging, slowly rotating cloud about one to four miles in diameter that drops down from the main storm cloud.

**In May 1999, a tornado cut a 40-mile path through the Bridge Creek and Moore areas just outside Oklahoma City. It completely destroyed more than 1,800 homes and damaged more than 2,500 other. The tornado also completely flattened a 600-unit apartment building and damaged more than 800 cars when it hit a car dealership. Thirty-eight people were killed in the disaster, and nearly 600 more were injured. Final damage estimates reached more than $1 billion.**

Simultaneous visual and radar sightings of wall clouds and other thunderstorm features have improved the interpretation of weather radar. This has enabled forecasters to recognize potentially dangerous storms more easily and issue more accurate and timely weather warnings. Moreover, chasers sometimes call in warnings from the field, long before a storm actually develops.

Photographs and videotapes of storms and the structural damage they cause, along with measurements of wind speed and pressure around storm systems, have been used to determine wind speeds around the vortex of a tornado. This information can be used to design safer buildings.

# The Best and Worst American Weather

What do you do when you know you don't live in the pleasant-weather capital of the world? Celebrate it!

At International Falls, Minnesota, a small town on the Rainy River, which separates Canada from the US, the hardy 9,000 or so residents have to brave January temperatures that average 1°F. Combined with brisk winds that average 8.9 mph, this results in a wind chill of about −17°F. That's cold enough to make even the toughest folks want to stay inside. And the winters at International Falls are long: the average temperature in November is 24.9°F, and it is not until April that average daily temperatures rise above freezing.

How do people here cope with this? Well, there's not much point pretending it doesn't exist, so they hunker down and take pride in being tough. International Falls holds an annual "Icebox Days" festival each January. One of the features of the festival is a 10K race called the "Freeze Yer Gizzard Blizzard" run.

There are many places in the US that have similar celebrations of their less-than-perfect weather. But the lack of perfection is not just a matter of cold. The US is geographically the third largest country in the world, and its weather is probably more varied than that of any other country. The huge US land mass stretches across all kinds of climates, from the high Arctic regions of Alaska to the near-tropical climate of Florida. It has deserts, open plains, high mountains, low-lying valleys, maritime coastal regions, and everything in between.

Although the American climate is mostly pleasant—but also stimulating, invigorating, and challenging—its unfavorable characteristics do create hardship. Both extreme cold and extreme heat are literally killers, but many other aspects of the weather can make life unpleasant, too. Lightning can be deadly. Hurricanes destroy lives and property. Fog, snow, and sleet make driving difficult. Too much rain soaks gardens and floods homes. Not enough rain kills crops. Too much humidity makes people feel lazy and tired. Air that is too dry chaps lips and leaves skin feeling like a parched mud cake. Too much sun can cause exhaustion,

The annual "Freeze Yer Gizzard Blizzard" 10K run in International Falls, Minnesota. *International Falls Chamber of Commerce*

sunstroke, and even heatstroke. Not enough sun can lead to depression. Too much wind makes outdoor sports like tennis, golf, and baseball awkward. Not enough wind makes sailing impossible.

And even in those areas of the country that enjoy good weather the majority of the time—like Santa Barbara, California, which has pleasantly warm days and comfortably cool evenings—there are other natural hazards like earthquakes and wildfires. No place is perfect.

So which places in the US have the best weather? Of course, it depends on personal preferences, but is there a perfect "10" climate? Who has the best and worst weather in the US, and where are the most depressing, hazardous, uncomfortable, and confining places, in terms of climate, in the US?

Knowledge of climate severity can be valuable to employers concerned about the timing of outdoor activities and workers' performance; to workers seeking fair and equitable remuneration for working outdoors or in isolated areas; to planners or designers of workplaces, residences, recreational areas, clothing, and equipment; and to persons who are retiring, or seeking havens from certain climate-related illnesses. Some stresses include extremes of hot or cold, wetness or dryness, and windiness; continuous darkness or daylight; prolonged or intense precipitation, fog, and restricted visibility; lightning and severe weather like thunderstorms, blowing snow, and freezing precipitation.

Since most relationships between climate and psychophysiological sensations are not well understood, one must rely principally on personal experience and observation in deciding what elements to include, and their relative importance, when assessing the quality of a climate. Considering the importance of comfort in our daily lives, which influences what we wear, how we feel, if and how we travel, and how effectively we work and play, this factor is important. Psychological state and hazardousness are complementary factors but are often of lesser importance than comfort because they are generally associated with less frequent and more ephemeral factors. Although weather restricts outdoor mobility, especially in winter, it is less significant as a year-round disruptive element than the three others.

## Discomfort

Winter is generally the most stressful time of the year in the US, though the height of summer can also pose challenges to those living in hot regions. The largest temperature contrasts in the US occur during winter, while relatively little contrast occurs during summer. (It has even been as hot as 100°F in Alaska in the summer.)

Winter discomfort depends on two main factors: the temperature and the wind speed. Together they provide a measure of the degree of coldness: wind chill. Wind chill is a recognized index of heat loss and cold injury for humans, combining the effects of low temperature and strong winds.

Based on January temperatures and wind speeds recorded by the National Climatic Data Center (NCDC) in various locations in the continental US, where would you want to be in January?

At the other extreme from cold is summer heat. Summer discomfort can be measured by heat and humidity. As expected, the hottest and most humid cities in July in the US are found in the South. Though the hottest cities tend to be in the Southwest, the most humid are in the Southeast.

The impact of humidity on how hot it actually feels is discussed in the first chapter. The US cities that are both hot and humid are among the least comfortable places to spend a summer. Air conditioning

## Chill Out: The 10 Coldest Places in January in the Continental US

| Location | Normal daily mean temperature in January (°F) | Normal daily minimum temperature in January (°F) |
|---|---|---|
| International Falls, Minnesota | 1.0 | −10.0 |
| Mt. Washington, New Hampshire | 3.9 | −4.6 |
| Fargo, North Dakota | 5.9 | −3.6 |
| Duluth, Minnesota | 7.0 | −2.2 |
| Saint Cloud, Minnesota | 8.1 | −2.4 |
| Williston, North Dakota | 8.9 | −1.8 |
| Caribou, Maine | 8.9 | −1.6 |
| Bismarck, North Dakota | 9.2 | −1.7 |
| Aberdeen, South Dakota | 10.1 | −0.6 |
| Glasgow, Montana | 10.6 | 1.2 |

## The 10 Windiest Places in January in the Continental US

| Location | Average wind speed in January (mph) | Normal daily mean temperature in January (°F) |
|---|---|---|
| Mt. Washington, New Hampshire | 46.3 | 3.9 |
| Blue Hill, Massachusetts | 17.4 | 25.3 |
| Casper, Wyoming | 16.3 | 22.4 |
| Cheyenne, Wyoming | 15.3 | 26.5 |
| Great Falls, Montana | 15.1 | 21.2 |
| Rochester, Minnesota | 14.4 | 11.5 |
| Buffalo, New York | 14.1 | 23.6 |
| Boston, Massachusetts | 13.8 | 28.6 |
| New York–La Guardia, New York | 13.8 | 31.3 |
| Dodge City, Kansas | 13.5 | 29.8 |

becomes a necessity and not a luxury. Clothes stick to us, our energy is zapped, and our bodies tell us to slow down. Nobody feels like going outside, and normal activities, like walking to the store to get a cold soda, become great chores.

The hottest cities aren't necessarily the most uncomfortable ones. In fact, of the cities in the following table, Waco is probably among the least comfortable in July. Yuma is a desert city, and a temperature of 106.6°F, with a relative humidity of 50, feels like about 138°F. In steamy Waco, a temperature of 96.8°F, with a relative humidity of 82, will feel like 142°F.

But for those who don't like prolonged heat, Waco is probably a better place to be than Yuma. Summers in Yuma are generally longer. In Yuma, seven months of the year include normal daily mean temperatures of over 70°F, while in Waco, there are just five months of that kind of heat.

## The 10 US Cities with the Highest Normal Daily Maximum July Temperatures

| Location | Normal daily maximum temperature (°F) | Average relative humidity, a.m. reading |
|---|---|---|
| Yuma, Arizona | 106.6 | 50 |
| Phoenix, Arizona | 105.9 | 44 |
| Las Vegas, Nevada | 105.9 | 28 |
| Tucson, Arizona | 99.4 | 56 |
| Fresno, California | 98.6 | 61 |
| Bakersfield, California | 98.5 | 48 |
| Redding, California | 98.3 | 58 |
| Bishop, California | 97.2 | 13 |
| Wichita Falls, Texas | 97.2 | 78 |
| Waco, Texas | 96.8 | 82 |

In general, summers in the northern parts of the country are more pleasant than in the South. Hot temperatures and high humidity are uncomfortable, and prolonged exposure can be dangerous. But winters in the North, and particularly the Northeast, can be difficult. It's not surprising, then, that so many Americans, like birds, flock south in the winter.

### Those Steamy July Days: The 10 US Cities with the Highest July Relative Humidity

| Location | Average relative humidity, a.m. reading | Normal daily maximum temperature (°F) | How hot it really feels |
|---|---|---|---|
| Port Arthur, Texas | 95 | 91.9 | 134 |
| Tallahassee, Florida | 94 | 91.3 | 128 |
| Jackson, Mississippi | 94 | 92.4 | 133 |
| Lake Charles, Louisiana | 94 | 90.8 | 127 |
| Gainesville, Florida | 94 | 90.7 | 127 |
| Quillayute, Washington | 94 | 68.2 | 137 |
| Corpus Christi, Texas | 93 | 93.3 | 135 |
| Houston, Texas | 93 | 92.7 | 132 |
| Meridian, Mississippi | 92 | 92.1 | 130 |
| Shreveport, Louisiana | 91 | 91.4 | 125 |

## Psychological State

People frequently blame weather for a host of psychological complaints, including fatigue, depression, irritability, sleep loss, lack of concentration, headaches, general nervousness, forgetfulness, photophobia (light intolerance), and chest and joint pain. Weather is also blamed for spells of "cabin fever" or general monotony among personnel in isolated postings.

Some of the climate elements that most influence one's psychological state are the amount of sunshine, the annual number of days with measurable precipitation, and the frequency of cloudy days.

Long periods of darkness, a characteristic of high-latitude winters, are known to be especially detrimental—adversely affecting moods, attitudes and behavior. Seasonal Affective Disorder (SAD) is a condition of depression that has been shown to relate to lack of sunshine. Most personnel who have lived for extended periods in the Arctic say the 24 hours of total darkness in winter is particularly debilitating. Some even suggest that the converse in summer, 24 hours of continuous light, is wearing at times. At the latitude of the Arctic Circle, 66.5° N, the polar night lasts 24 hours. The length of the polar night increases nonuniformly with latitude, so that at the North Pole, it is six months long.

Sunshine has important physiological and psychological implications. Clear, sunny weather, especially at the end of a long spell of overcast, can be mentally uplifting. The sunniest region of the US is the Southwest, not Florida, although that state is known as "The Sunshine State." Even Key West, the sunniest city in Florida, just ties for 10th place.

## The 10 Sunniest Cities in the Continental US

| Location | Average annual percentage of possible sunshine |
|---|---|
| Yuma, Arizona | 90 |
| Redding, California | 88 |
| Phoenix, Arizona | 85 |
| Tucson, Arizona | 85 |
| Las Vegas, Nevada | 85 |
| Dallas–Ft. Worth, Texas | 84 |
| Fresno, California | 79 |
| Reno, Nevada | 79 |
| Flagstaff, Arizona | 78 |
| Pueblo, Colorado, and Key West, Florida | 76 |

## The 10 Cloudiest Cities in the Continental US

| Location | Average number of cloudy days per year |
|---|---|
| Quillayute, Washington | 239 |
| Astoria, Oregon | 239 |
| Olympia, Washington | 228 |
| Seattle Sea-Tac Airport, Washington | 226 |
| Portland, Oregon | 222 |
| Missoula, Montana | 214 |
| Elkins, West Virginia | 212 |
| Binghampton, New York | 212 |
| Beckley, West Virginia | 210 |
| Sault Ste. Marie, Missouri | 209 |

Frequent wet or foggy days can be demoralizing. This is especially true for northern residents, who, after enduring a long winter, feel discouraged if precipitation occurs frequently during the warmest time of the year. This feeling is somewhat similar to that experienced by those who spend the winter months in total darkness. Among the cloudiest and rainiest cities in the US are Quillayute, Washington, and Astoria, Oregon.

### When Will the Rain Stop? The 10 Cities in the Continental US with the Most Days of Precipitation

| Location | Mean number of days per year with measurable precipitation |
|---|---|
| Quillayute, Washington | 208 |
| Astoria, Oregon | 191 |
| Elkins, West Virginia | 171 |
| Buffalo, New York | 169 |
| Marquette, Missouri | 166 |
| Sault-Ste. Marie, Missouri | 166 |
| Erie, Pennsylvania | 164 |
| Olympia, Washington | 162 |
| Caribou, Maine | 161 |
| Binghampton, New York | 161 |

## Hazardousness

Elements of climate either singly or together can produce widespread injury and death and bring about considerable damage to property and the environment. Obvious examples are floods and blizzards, which may seriously disrupt entire communities. Extreme wind chill in winter and excessive heat and humidity in summer are also hazardous.

Weather hazardousness can be evaluated by considering the average winter snowfall and the frequencies of other elements like strong winds, hurricanes, and tornadoes. These phenomena can cause a host of problems, including possible death, injuries, missed social and business events, delayed services, and other privations. Heavy snowfalls create many personal hardships because of confinement and possible shortages of food and fuel. In populated areas, heavy snowfalls are usually accompanied by a spate of heart attacks due to overexertion. Blowing and

## The 10 Snowiest Cities in the Continental US

| Location | Annual average snowfall (inches) |
|---|---|
| Marquette, Missouri | 130.6 |
| Sault Ste. Marie, Missouri | 117.1 |
| Syracuse, New York | 114.7 |
| Caribou, Maine | 110.7 |
| Lander, Wyoming | 102.2 |
| Muskegon, Missouri | 97.9 |
| Buffalo, New York | 91.8 |
| Erie, Pennsylvania | 86.5 |
| Alpena, Missouri | 85.7 |
| Binghampton, New York | 82.8 |

## The 10 States with the Highest Average Number of Tornadoes per Year, 1950–95

| State | Annual average number of tornadoes |
|---|---|
| Texas | 124 |
| Oklahoma | 52 |
| Kansas | 47 |
| Florida | 46 |
| Nebraska | 37 |
| Iowa | 31 |
| Illinois | 26 |
| Missouri | 26 |
| Colorado | 25 |
| Louisiana | 25 |

drifting snow create dangerous outdoor travel conditions, stranding people in perilous situations. In fact, all outdoor activity becomes extremely hazardous if near-zero visibilities in blowing snow are combined with high wind chill. Under such conditions, farmers have become lost and died of exposure while attempting to walk from the barn to the house.

But if you want to escape snow and cold by moving to a warmer climate, you may face other hazards. For example, if you are afraid of getting hit by lightning, don't move to Florida. Florida has had over 350 lightning deaths since 1959—by far the highest of any state, about double the number of deaths in North Carolina and Texas, the runners-up.

Of these snowy locations, Buffalo is the windiest—its January wind blows an average of 14.1 mph. Wind has blasted the city at speeds as high as 91 mph. On days like that, everyone wants to stay inside.

## Outdoor Mobility

Our ability to move about, to travel to work, school, and shopping is restricted by adverse weather. Three sub-factors have been identified as restricting outdoor travel and access: total snowfall, freezing precipitation, and limited visibility. In the US, snowfall must be taken into account for assessing the ease of outdoor movement on foot and by vehicle. Freezing precipitation restricts all forms of transportation, from walking to flying. In addition, it frequently plays havoc with communication, owing to downed telephone and power lines. Visibility reduced by fog, snow, or rain also limits our ability to move about and travel.

Buffalo, New York, has often been the victim of bad weather jokes. Unfortunately, the jokes are based on solid statistics. If you care about good weather, Buffalo has simply one of the toughest climates in the continental US. Besides having one of the snowiest climates, it's also very windy. Winters are long and cold. By the end of an average November, 11.3 inches of snow have already fallen, and over three inches usually fall in April.

Unfortunately, there are plenty of other unpleasant features about Buffalo's climate: it receives only 49 percent of possible sunshine, it is cloudy 208 days of the year, its summers are humid, and it rains or snows on 169 days of each year. It has clear skies just 54 days each year. There is, however, at least one good thing about Buffalo: tornadoes and hurricanes are rare. (It also boasts a low crime rate.)

Among the best places to live? If you like sunshine, go to the Southwest—places like Phoenix, Yuma, Tucson, and Las Vegas are sunny and hot and have low relative humidities. But summer temperatures can skyrocket and leave those unlucky people without air conditioning in danger.

The Southeast also has a warm climate year-round, but Florida, the

Gulf Coast, and the Carolinas are prone to hurricanes (not to mention lightning), and Texas and Oklahoma are likely to get tornadoes.

The honor of having the absolute worst climate in the continental US goes to Mount Washington, New Hampshire. It is the highest peak in the northeastern US, and is among the cloudiest, chilliest, windiest, and snowiest places in the lower 48 states where weather conditions are recorded by the NCDC. In fact, one of the highest wind speeds ever recorded was observed at Mount Washington: the wind whipped the mountain at 231 mph.

Luckily, nobody lives there!

## When the Weather's Not Sporting

Nobody thought the boys from Florida could beat both the tough northern team and the tough northern weather.

"I want the Marlins to realize," a confident Cleveland Indians fan told the Associated Press, "how cold it is here and that they don't have a chance."

Another Indians fan held up a sign: "Not in Our Igloo."

It was the 1997 World Series. The Florida Marlins faced the Cleveland Indians. The first two games were played in weather 70°F to 80°F in Florida, but then, somewhere on the flight from Florida to Cleveland, the temperature fell by at least 40°F. When the two teams reached Cleveland, the games were forecast to be played in near-freezing temperatures and snow flurries. Luckily the Marlins had packed their winter parkas on the plane. And the Marlins' equipment manager told the *Seattle Times*, when asked what he had brought for his team in case of the weather: "Warm thoughts."

**Only once in its 96-year history has a World Series game been postponed because of cold weather. It was during the very first Series, in 1903, in Pittsburgh.**

It snowed during Game 3, which was the first time snow had fallen during a World

Series game since 1979, when the Orioles hosted Pittsburgh in Baltimore.

**The longest rain delay in World Series history was in 1911. Game 4 in Philadelphia was delayed six days, from October 18–24.**

The Cleveland snow should have favored the home team, but the Indians must have had too much of that balmy southern air. Florida ended up with 17 walks and won the game, beating the Indians 14–11.

Game 4 on October 22 was a chilly 38°F, the coldest World Series game on record. The wind chill at times was as low as 18°F. Some umpires wore ski masks to keep warm. Players wore gloves on the sidelines, and bundled up in sweatshirts and jackets during practice. Heaters blasted warm air into the dugouts.

"It was the coldest weather I've ever pitched in," Cleveland pitcher Jaret Wright told *USA Today*. Just before the game started, as players warmed up (if that was possible), stadium loudspeakers blasted out the song "Winter Wonderland." Florida finally succumbed to the cold and lost that game 10–3. Marlins manager Jim Leyland blamed it on the weather.

**Weather isn't the only natural hazard that has affected the World Series. A powerful earthquake in San Francisco just before Game 3 of the 1989 World Series between the San Francisco Giants and the Oakland Athletics delayed the Series for 10 days.**

"We have been a little sloppy. I think the conditions have something to do with that," Leyland told the Associated Press.

Game 5 was warmer, and Game 6 was played in Florida. When it was all over, the Marlins had won the 1997 World Series 4–3. They had shown that even warm-water fish like Marlins can weather the weather.

North Americans love sports, but when the weather strikes, we're often left out in the cold. From cold snaps that make baseball games miserable, to lightning that terrorizes golf players, to searing heat and humidity, which can make even the hardiest athletes want to wilt, North Americans have to cope with all kinds of weather when it comes to both recreational and professional sports.

Outdoor sports are particularly vulnerable to the wrath of the weather. Indoor games like hockey, gymnastics, and skating are less likely to be canceled because of bad weather, but even indoor events are affected if transportation is hindered. A baseball game at the covered Houston Astrodome was "rained out" on June 15, 1976, when heavy rains caused dangerous flash floods in the streets of Houston and prevented fans from getting to the stadium. And a hockey game on February 28, 1959, in an arena in southern Ontario ended in tragedy when the roof and walls caved in under the weight of the snow. Eight people died.

Wind affects cyclists, runners, tennis players, football players, and sailors. Too much wind sends balls flying in the wrong direction, or worse, can topple sailboats, motorcycles, and bicycles. Skiers also have to be wary of high winds. On March 27, 1986, wind gusts of up to 100 mph closed a ski hill near Juneau, Alaska. Just before Christmas in 1964, winds of 100 mph forced the rescue of 125 skiers from an Aspen chairlift.

Lightning has threatened and sometimes attacked golfers on courses all over the country. Bobby Nichols, Lee Trevino, Steve Ballesteros, and Tony Jacklin have all been hit by lightning. Two golfers on a West Virginia golf course were hit by lightning in 1980, and one was hurled 20 feet.

Golfers aren't the only ones at risk from lightning. On July 13, 1990, a group of hikers on California's Mount Whitney found themselves caught in a thunderstorm, and sought shelter in a metal-roofed stone building near the summit. Lightning struck the stone hut, and all 13 hikers felt the pain of the shock. The unlucky one was 26-year-old Matthew Nordbrock, who had been wearing metal-rimmed glasses at the time the bolt struck. Despite his companions' attempts to resuscitate him, he died.

On September 14, 1984, 26 people were hurt on a Pennsylvania soccer field after being hit by lightning. One person died a few days later in the hospital. A football player at practice in Silver Spring, Maryland, was hit on the head by lightning on September 19, 1992. His helmet was damaged, but, amazingly, he survived. In Camden, Maine, a ski lift was hit by lightning in December 1989. Spectators at a July 1987 basketball game in Illinois were leaving the game when a bolt of lightning struck a nearby tree and metal fence; 30 people were hurt. And on April 7, 1984, lightning hit a soccer coach and his players while they were putting up a net by standing on an aluminum ladder in Texas. CPR saved his life.

Rain, fog, and snow are also common reasons for the cancellation or postponement of sporting events. During the 1998 Winter Olympics at Nagano, Japan, the men's 10K biathlon was halted due to heavy snow and fog. At the time it was canceled, almost a quarter of the 73 competitors had already completed the event. It was rescheduled for the following day. Olympic officials said it was the first time an international biathlon event had been canceled since 1972.

**Thick black smoke, fluffy ashes, and strained community resources resulting from Florida's extensive forest fires forced the cancellation of NASCAR races in Daytona on July 1, 1998.**

If you ask an alpine skier if there's ever too much snow for skiing, most would say no. But at the 1998 Nagano Olympics, the men's giant slalom event was postponed when as much as four feet of snow blanketed Mount Higashidate. It was the sixth day at the Nagano Games that alpine races were postponed due to the weather. Bad weather also forced alpine skiing event cancellations at the 1992 Albertville Olympics.

**There are over 20 cities in North America that have built covered stadiums to escape from inclement weather. And some cities, like Toronto, Canada, have stadiums with retractable roofs.**

Hail can pummel players and spectators and leave them running for cover. On September 22, 1993, golf-ball-sized hail hurt 12 high school football players in Kansas.

Tornadoes are often threatening to sports, but that doesn't mean that organizers always pay attention to the dangers. The Tornado Project reported that a baseball game on April 8, 1998, between the Birmingham Barons and the Carolina Mudcats was played through rain, lightning, and even the sounding of tornado sirens. The umps didn't call off the game until they learned that they were in the path of an approaching tornado. Just over an hour later, they resumed the game. They were among the lucky ones: 34 people died in the outbreak of tornadoes in Alabama that day.

In a bizarre event in Canada on July 1, 1985, a 26-year-old man was killed when he tangled with a mini-tornado while para-sailing on a lake in Alberta. The victim was wearing a special para-sail, similar to a regular parachute, and was being pulled behind a boat just above the surface of the water. Although the winds were calm at the time, a whirlwind no wider than 15 yards suddenly appeared from the west and enveloped the man. He didn't see it coming until it hit. The whirlwind lifted the man about 60 yards into the air, snapping his 90-yard-long nylon tow rope, and then dropped him in a field several hundred feet east of the lake, where he struck a barbed-wire fence before landing. He died on impact.

**The PGA Tour was plagued with bad weather throughout 1998. Some events were rained out, some were canceled because of strong winds or thick fog, and by June, 12 events had already been canceled due to bad weather.**

Heat and humidity take a heavy toll on the human body, but often the show must go on. At the final round of the 1964 US Open golf tournament in Washington, DC, the temperature was 92°F and the humidity was 85, resulting in an apparent temperature of 123°F. Luckily the golfers were permitted to cool down with ice. Horses and athletes beat the heat at Atlanta's 1996 Summer Olympics by relaxing under misting fans—which can cool the air by as much as 25°F—and by drinking plenty of ice water.

**Snow fell at a rate of two inches per hour during the NFL championship game, dubbed the "Blizzard Bowl," in Philadelphia on December 19, 1948.**

Can the weather ever be too good for playing sports? Well, on July 1, 1974, a Montreal Expos game was delayed because the glare from the sun was too strong.

# Weather Observing and Forecasting

# The National Weather Service

Today, people expect to know exactly what the weather is going to be like. We want to know how warm it will be, how cold it will be, whether it will rain or snow, and how windy it will get. And we don't want to know just the weather outlook for one day; we expect long-range forecasts for several days. Some people even want to know what the weather will be like many months in the future.

**Weather forecasting was not always a wise thing to do. In 17th century England, the stakes were high. One law read: "Death by burning shall be the punishment for the practice of weather forecasting."**

The National Weather Service (NWS) is the main source of weather forecasts for the US. The NWS also provides information on severe weather watches and warnings. It processes some 400,000 meteorological bulletins each day.

Armed with state-of-the-art technology such as Doppler radar, the Automated Surface Observing System, Geostationary Operational Environmental satellites, and sophisticated computer forecasting models, the NWS helps meteorologists to provide us with much of that information we demand.

But even though we didn't always have advanced technology, people have a long history of watching, recording, predicting, and analyzing the weather.

**In 1894, the first observations of air temperature aloft were made by William Eddy. He used kites to send a thermometer into the sky.**

The earliest weather records kept in the US date back to 1644. In that year, the Reverend John Campanius started a collection of weather records at Swedes' Fort, near Wilmington, Delaware.

George Washington was an avid weather observer, and Thomas Jefferson was even more dedicated to watching the climate. While he was busy writing the Declaration of Independence, Jefferson bought his first thermometer. He got his first barometer a few days after Independence Day, on July 7, 1776. He was known for making two weather observations each day, one in the early morning and one in the afternoon. He ensured that Lewis and Clark were equipped with weather instruments for their expedition.

**Charles Lindbergh checked with the Weather Bureau before his transatlantic flight from New York to Paris in 1927. The Weather Bureau had predicted fog and rain over the Atlantic; it was right, but Lindbergh had departed before hearing the final forecast.**

These steps seem simple, but they were pioneering at the time. It was not until the first half of the 19th century that small weather observing networks developed in the US. Some of the observers were physicians; in 1814, Surgeon General James Tilton ordered that field doctors keep records of weather as well as human health.

Joseph Henry, an early director of the Smithsonian Institution, was one of the first organizers of weather forecasting in the US. The opening of the telegraph system to the public in 1845 helped communications immensely, and in 1849, Henry was able to convince the telegraph companies to provide free air time so that daily meteorological reports could be sent by volunteers to the Smithsonian. The following year, Henry started charting maps from the data he received. Comprehensive information was published and diffused, setting the stage for the Smithsonian's mandate. There were about 500 weather observation stations by 1860, when the Civil War broke out.

**Weather forecasts, which called for favorable winds and tides, prompted the decision to invade Normandy on June 6, 1944.**

In 1868, Professor Cleveland Abbe, an astronomer by training, drew up the blueprints for a national weather observation system. Abbe was the director of an observatory in Cincinnati, and he compiled weather data and charts, but he believed a formal system of observations on a broader scale could lead to accurate forecasting.

On February 9, 1870, President Ulysses S. Grant approved the establishment of a national weather service. It was originally under the jurisdiction of the Secretary of War, assigned to the Signal Service Corps. What was this national weather service's original name? The Division of Telegrams and Reports for the Benefit of Commerce.

**During World War II, restrictions were imposed on weather broadcasts in the US because they "could aid the enemy."**

On November 1, 1870, the first systemized weather observations were made by Signal Service agents at 22 stations and sent to Washington. By 1878, there were 284 observation sites. However, according to the NWS, "Little meteorological science was used to make weather forecasts during those early days. Instead, weather that occurred in one location was assumed to move into the next area downstream."

The weather service was transferred from military to civilian hands in 1890. It was renamed the US Weather Bureau, under the jurisdiction of the Department of Agriculture. (It remained with Agriculture until 1940.) During this period, its budget, staff, and mandate were enlarged. It became responsible for issuing flood warnings to people (the worst flood in US history at Johnstown, Pennsylvania, occurred in May 1889). It started publishing the first daily weather map of Washington in 1895.

**NWS warnings, watches, and forecasts are broadcast 24 hours a day on the National Oceanic and Atmospheric Administration's Weather Radio network of over 480 stations in the US. Weather service scientists hope that eventually weather broadcasts will reach 95 percent of the population.**

In 1898, President McKinley ordered that the Weather Bureau set up a hurricane warning system in the Caribbean. Following the deadliest hurricane in US history, which hit Galveston, Texas, in 1900 and killed over 6,000 people, the Weather Bureau began making three-day hurricane forecasts. By 1935, a full hurricane warning system was in place.

In 1902, Cunard ships at sea began to receive Weather Bureau forecasts by wireless telegraph. In 1903, the Wright brothers consulted with the Weather Bureau to prepare for their flight. By 1910, the Weather

Bureau was sending out weekly forecasts for agricultural planners. Fire weather forecasts started to be issued in 1916.

The advent of aviation demanded more sophisticated means of tracking the weather, and it also provided an unprecedented tool for observing it. In 1918, the Weather Bureau began to send out forecasts to military and air-mail flights. By 1931, the Weather Bureau was operating regular observation flights, marking the end of the era of kites in weather observation. The close links between aviation and meteorology led President Roosevelt to transfer the Weather Bureau to the Department of Commerce in 1940.

The development of computers, radar, and satellites in the 1950s and onward has given the NWS even greater means for analyzing and predicting weather patterns. In 1977, the last US weather observation ship was sent back to port, its job now done entirely by satellites.

Since 1970, what was once the Weather Bureau has been called the National Weather Service. At that time it was integrated into the newly formed National Oceanic and Atmospheric Administration.

Today, the NWS provides an invaluable service to all Americans. It is one of the most utilized public services in the country, with an incredibly low cost of between $1 and $2 per taxpayer per year. The NWS advises Americans when there's a hurricane or

**Advance notice of weather events like hurricanes, tornadoes, and blizzards gives people time to prepare, and to evacuate if necessary. Though forecasting is an imperfect science, and weather will often change its course at the last minute, forecasting saves lives since it allows disaster management officials to call for evacuation. During Hurricane Andrew in August 1992, the death toll was remarkably low because most residents had been evacuated by the time Andrew slammed into the land. The warning time for a tornado is much shorter, though. With modern technology, the NWS can sometimes provide as much as 12 minutes of lead time during a twister—enough time for potential victims to get to storm shelters.**

tornado coming, it helps farmers plan their crops, it provides pilots with information that helps them decide where, when, and sometimes whether to fly, and it allows people to make all kinds of personal and professional choices that relate to the weather.

The NWS is currently undergoing a major modernization and restructuring program. Incorporating new technologies is part of that program, and improved forecasting methods will no doubt follow.

Throughout its long history, the National Weather Service has been dedicated to a simple but difficult principle: although we can never control the weather, we can cope better with it if we know what to expect.

## "And Now for the Weather . . ."

Television weather presenting has changed dramatically since the no-frills days when the tools of the trade were chalk and blackboards: high-tech then meant magic markers and magnetic smiling suns. Today, computer graphics have revolutionized the industry. Weathercasters no longer have to draw their charts by hand or use stick-on symbols or marker boards. Instead, all television stations have an array of commercially available animation and data-presentation software to enhance appearance and add interest and value to the weather. Stations woo viewers by boasting the latest in weather tools and electronic gadgetry: three-dimensional clouds tracked by satellites, color radar imagery, and a set replete with banks of monitors and instruments that look like they came from the Starship Enterprise.

Most television presenters use a special effect called a chroma-key panel. The person pretends to point at a weather map that is really a large blank wall—an area of green or blue painted plywood or felt board. The weather map or graphic exists in the computer inside the control booth. The camera senses the chroma-key color and electronically superimposes the graphic on the wall. Then right on cue, the weathercaster clicks through a fast-paced assortment of colorful and informative maps and graphics. Hosts know where to point by looking out the corner of their eyes at TV monitors on each side of the wall or on the camera in front. Skilled presenters can create the illusion that they are touching real maps.

Technology aside, forecasts are still only as accurate as the information that goes into them. Weathercasters are not bound to follow the official weather service forecast. Many make their own interpretations, often by just altering the timing of approaching weather or by changing rain to snow.

An effective television weathercast depends largely on the presenter. A good weather broadcaster must be a jack of all trades—a forecaster, computer technician, educator, journalist, psychologist, and entertainer. Among the necessary qualities for wannabe weathercasters are personality, enthusiasm, authority, and a sense of humor. Above all, it requires someone with good communication skills to present relatively dry scientific information in an engaging way that gives the public what they need to know. At the same time, because weather is much more than rain or shine, they must be able to explain succinctly and clearly how the weather systems behave and why.

**A woman in Israel sued a television station and its popular weathercaster in small-claims court for the equivalent of about $1,000 after he predicted sun for a day it rained. The woman claimed his forecast caused her to leave home lightly dressed. The unforeseen shower messed up her hair, gave her a case of the flu, which made her miss four days of work, and caused her mental distress. More insulting than the final compensation sought was the woman's demand that the forecaster apologize for his error. The court dismissed the case.**

An estimated 50 percent or more of television weathercasters in the United States are professional meteorologists or have a science degree. Unlike news readers, weather personalities can put their own personal stamp on their telecasts. Some read poems or riddles on air, while many show viewers' photographs or children's drawings. Others have set up a network of volunteer weather spotters or spies who call the station's weather hotline daily. Some weathercasters present the weather outdoors on the street or from the station's roof or patio. Others deliver news of the weather outside from a remote site, sometimes tying it in to a charity event like a TV auction, telethon, or celebration. Local boosterism and public-service announcements fit nicely into the weather segment of the newshour.

Weather shticks and gimmicks abound, as one or two "hippy dippy" weathercasters per market area mix silly antics with isobars and wear outrageous costumes while explaining wind chill. Forecasters have delivered weather reports to live turkeys at Thanksgiving, worn barrels at income-tax deadlines, and routinely paraded pets and ridiculous mascots. In Phoenix, Arizona, a French poodle named Puffy Little Cloud would appear wearing an outfit appropriate to the weather—doggy raincoat on wet days and a hand-knitted sweater on chilly ones. No wonder one writer once remarked that "most weathercasters resemble a lounge comedian who did a semester at MIT."

**Weather launched the media careers of Diane Sawyer, Pat Sajak, Dick van Dyke, Raquel Welch, Dan Rather, and David Letterman.**

The impetus behind such antics may well be the nature of weather—90 percent of the days aren't exciting or eventful. How many fresh ways can you say "sunny and cold"? For three decades, a Detroit weathercaster named Sonny Elliot used to entertain his audience by inventing new weather words. It became his trademark—a snowy and breezy day became "sneezy"; showers and windy made for "shindy" conditions; and fog and drizzle produced "frizzle."

All television weathercasters are enthusiastic when severe weather threatens. Of course, having the lead story with frequent updates throughout the newscast makes them feel important and needed. And they're right! It's hard to think of a news or sports story that is as crucial to the public as a promptly aired weather advisory.

Unfortunately, weathercasters are too often convenient targets for sarcastic remarks, "Thanks for that *great* day!" or "How come you weather guys never get it right!" In the weather business, you can't please all of the people all of the time. You don't even try! The exasperated farmer who did not get a shower when all those around him did; the restaurateur with the outdoor patio, irate that rain is mentioned yet

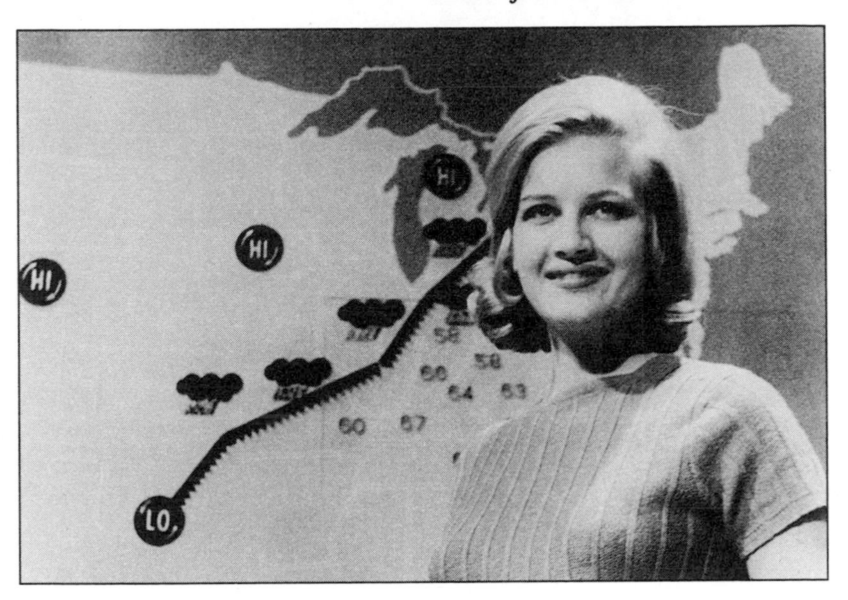


*Good Morning America* co-anchor Diane Sawyer began her broadcasting career in Louisville, Kentucky, as a weathercaster. *ABC Television*

again; and the anxious skier annoyed by winter's late arrival. A Canadian weathercaster named Mike Roberts has been hanged in effigy for too much rain and too much sun. One woman angrily confronted him for forecasting isolated showers. She thought it meant ice showers and worried every time he mentioned them. And a trailer-camp operator threatened to sue a weathercaster because he predicted rainy weather when the day turned out beautiful and sunny. Believe it or not, weathercasters have even been accused of creating the weather.

One common complaint, justified at times, is the tendency of weathercasters to pack too many weather details into four minutes, droning on about fronts and troughs, dew point and barometric pressure. This can be particularly annoying when all you want to know is whether it's going to rain tomorrow. Also, why must they show a map with temperatures from 100 nearby locales that are no more than a couple of degrees apart?

As for weathercasters, one of their grievances is snoozing anchors. It's bad enough being the brunt of stale one-liners every night and having the co-hosts and crew vacate the set when the weather segment comes up. But it is annoying to hear the anchor say, "Great, looking forward to a weekend of fine weather," when you have just forecast two days of overcast skies with showers.

Television weather presenters should demand more time from program managers and more respect from both their colleagues and the public. Audience studies in every major market consistently show that the television weathercast is one of the main reasons people tune into local news. Ratings for weathercasts are way ahead of sports and even above the top news story.

## Meteorological Mumbo Jumbo

What could be so perfectly comprehensible as: sunny with a few cloudy periods; windy with the chance of an afternoon shower; warmer, with a high of 77?

However, almost all the everyday words that the weather forecaster uses are at times vague, complex, or ambiguous. *Bright, clear, intermittent, variable cloudiness*, and *likely* all give rise to misinterpretation.

A weather forecast, to be useful, has to be clear, specific, and impersonal, and should be worded in such a way that millions of people instantly understand it. People prefer the weather presented in conversational terms. They are befuddled by terms such as *trowal, trajectory*, and *tropopause.*

And it should be concise: people can absorb and remember only so much information at one time. Most people simply want to know whether it is going to rain or snow today, not where the cold front was three days ago. One survey found that more than half of those who watch a weather report on television cannot recall where or how much it is expected to rain, even seconds after the forecast has been given. We may all hear the same forecast, but we often come away with a different message. Mind you, some weathercasters' antics, jabber, visual wizardry, and gimmicks can obscure the essential message.

In the end, the value of an accurate forecast hinges on how well it is understood. But has the average person the same conception of weather and weather terms as the meteorologist? Qualifiers such as *intermittent, variable, occasional,* and *scattered* have unmistakable meaning to forecasters, but they may convey something quite different to the public. *Fair* weather may mean clear skies to one person, but merely the absence of rain to another. Even though a forecast is issued, and later verified, it is often deemed either a bust by the general public, or even worse, it is misinterpreted. The consequences can range from merely bothersome (you plan a picnic based on what seemed like a good forecast, but it rains) to life threatening (being caught on a golf course in a thunderstorm). Such incidents hurt the weather forecaster's credibility. Besides being the most maligned, they become the most misquoted of human beings.

Sometimes people seem to remember only what they want to hear. *Partly sunny* may be recollected as *sunny*. And a forecast that calls for a "30 percent possibility of rain" may be seen as misinformation if it doesn't rain, if the element of probability is overlooked.

Several phrases used in official forecasts are often confused or misunderstood. Do you know what a 70 percent probability of rain means? In a national survey, only 10 percent of respondents knew it meant that on 7 days out of 10 with a similar weather pattern, a given place within the forecast area (your house or the airport, for example) will receive measurable rainfall. In simpler terms, a probability of precipitation forecast (POP) is just what it sounds like—an expression of the forecaster's confidence that there will be rain or snow. If there is a good chance, the forecaster may say 80 percent; if dry weather is expected, the POP may be 20 percent. Officially, POP refers to the probability of .01 inches or greater of precipitation at any given point in the forecast area over a specified time period.

Meteorologists do worry about whether the public fully understands their forecasts. Studies over the past 50 years indicate that most people understand most of the terms in daily forecasts. However, differences do exist between the meteorological meanings of some expressions and how people perceive them. To reduce the confusion, a few words have been abandoned, existing ones revised, and new ones formulated.

Fortunately for the public, much of the meteorologist's specialized vocabulary never makes it into official forecasts. Visit a weather office and you'll hear conversations peppered with such technical terms as *cold low, dry tongue*, and *helicity*. Then there are everyday words with special meanings, such as *waves*—as in cold wave—or *front* in the sense of a surface where masses of warm and cold air come into conflict. Over the years, a variety of other meteorological mumbo jumbo has crept into the forecaster's lingo, if not into the official public forecast, such as *scuds, bombs* and *busts, dry lines, moist tongues*, and *hot boxes*, to name a few. Meteorology is not without word play—*toe-lapper, cirrusnirrus*, and *severe clear*—and abbreviations, including *TTTC* for weather that is "too tough to call" or *BICO*, for "Baby, it's cold outside." On the other hand, the origin of some words is no mystery, although the context may be: Alberta clipper, Colorado hooker, and Texas gully washer.

Some weather words really don't mean what they may seem to. For example, *cloud cover* is really sky cover, *zero visibility* includes visibility up to a distance of below 100 yards, *probability* doesn't mean probable, *wet hail* is not hail, and *meteorology* has nothing to do with meteors.

Some words have been retired from use in public forecasts. For example, you no longer see the expression *unsettled weather* in an official public forecast. At one time, it meant brief periods of fine weather alternating with periods of rainy, cloudy, or stormy weather, a kind of variable weather. As retired meteorologist Reuben Hornstein explained, "The public joked that a forecast of unsettled weather might mean either that the weather was going to be extremely changeable, or that the forecaster was unsettled about what to forecast and took refuge in that rather vague and indeterminate wording."

Another word not appropriate in official forecasts is *fair*, once used to describe days with sunny intervals, or warm and pleasant days without sunshine. Following much public criticism that it was vague, the term was dropped from the forecaster's slate of official weather words. Likewise, *cloudy with sunny periods* and *sunny with cloudy periods* are not permitted because few people understand the shades of meaning between them.

Subjective comments such as *fine* are also considered inappropriate—sunny weather may be great news for resort owners or construction companies, but it could be a disaster for farmers hoping for drought-ending rains.

Some widely used words are not part of the meteorologist's vernacular. For example, Canadian weather officials never adopted the term *sleet*, which was once used in the United States to describe solid grains of ice formed by raindrops freezing before contact with the ground. They bounce and make a sound upon impact. In Canada, they call them ice pellets. In the United Kingdom, sleet means something entirely different—rain and snow falling together, or snow that melts as it meets the ground.

Even expressions coined by meteorologists to make a weather concept more comprehensible are open to misinterpretation. One example is *wind chill*, used to describe what cold weather feels like at various combinations of low temperatures and high winds. The problem is that some people confuse it with the actual temperature. Moreover, the formula used to calculate wind chill doesn't take into account factors such as incoming and outgoing radiation, sunshine, humidity and human metabolic rate—important elements in how acutely individuals feel the cold.

Some weather expressions are used only in specific regions. *Chinooks* are the dry, warm winds from the west that occur principally in the Rocky Mountain states; and on the eastern seaboard, a *Nor'easter* is the cold winter wind that blasts the east coast and results in heavy snow and sometimes blizzard conditions. To better serve regional needs, local forecasters have some flexibility in their choice of terminology and forecasting criteria. For example, a weather center in Washington, DC, might issue a "heavy snow" warning, which could mean a couple of inches are on the way to the national capital. The same amount of snow in Buffalo, New York, would hardly make the headlines.

The following is a sampling of meteorologists' jargon:

- *Alberta clipper*: a rapidly moving storm that streaks out of the Canadian Prairies into the Great Lakes and New England or areas farther south, often leaving a dusting of snow
- *bomb*: unforeseen and not forecast weather; also called a gremlin
- *cirrusnirrus*: thin, wispy, high cirrus cloud that forewarns of an approaching storm
- *cold low*: a weather system that consists of a low-pressure area

riding above a mass of cold air at the earth's surface, bounded by slowly swirling bands of cloud and precipitation in the upper atmosphere

- *Colorado hooker*: a storm that originates over the eastern plains of Colorado and moves across the Central Plains and Mississippi Valley; the storm taps moisture from the Gulf of Mexico and pulls cold air down from Canada; its hooked, or curved, path sometimes brings it into southern Canada, where it dumps heavy snow
- *dead clouds*: cumulus clouds that blot out the sun but are usually incapable of generating precipitation
- *helicity*: a measure of the potential of a small area of the atmosphere to spin rapidly, forming a tornado; expressed as a number, it is used to describe the danger of tornadoes forming within individual thunderstorms
- *hot box*: a localized storm area squared off on a weather map for which a meteorologist is likely to issue severe storm warnings; helicity is often closely monitored within hot boxes
- *severe clear*: not a cloud in the sky
- *suckerhole*: a brief period of clear weather that lasts until just after the good weather forecast has been issued or until a pilot flying without instruments takes off; only a sucker amends the forecast to reflect the good weather, and only a sucker flies without instruments when such conditions occur
- *Texas gully washer*: rain intense enough to flood gullies
- *TTTC*: weather that is "too tough to call"

# A Wild Winter Is on the Way—Maybe

When farmer Ed Burt spotted a wasps' nest on the cross-arm of a power pole in the summer of 1992, he was worried.

Burt is a homegrown weather prophet who believes nature's creatures can foretell weather conditions. The higher off the ground wasps, bees and hornets build their homes, he maintains, the deeper the winter snow will be. And if they build in sheltered areas such as barns and woodsheds, expect a long, harsh winter. In 1993, Burt found three wasps' nests close to the ground. "Sure enough, there was little snow that year," he says. However, he admits the power pole nest signified nothing more than wasps with a penchant for penthouse living; snowfall that winter was below normal.

Gordon Restoule is another believer. He operates a tourist camp on the Dokis Indian Reserve in Ontario, Canada. For him, it's the thickness of mud that beavers plaster on their lodges that reveals the misery level of the coming winter: the more mud, the worse the winter. Geese flying south early and high up promise a rough winter too, he says. Restoule claims 80 percent accuracy, but acknowledges he blew it the winter of 1993–94 when he predicted a mild winter and it turned out to be the coldest in nearly 75 years.

In the West, Roy Robertson, a retired farmer and former trapper who now works in season as a greens keeper at a golf course, keeps his eye on woodlot denizens. If squirrels are frantically storing cones and mushrooms, look for a cold, snowy winter, he declares. Expect a warmer one with little snow if beavers are late starting their winter feed beds, or if elk and moose are late in mating. Mind you, says the soft-spoken former trapper, "Nature throws the odd curve! I missed the heavy snow last winter, but it didn't matter much, because I spent the season in Arizona."

Homey weather wisdom has been the basis for weather predictions for thousands of years. People whose livelihoods depended on the environment observed the regular patterns of nature: the changing of the seasons, the sun's daily progress, the moon's phases, animal hibernations, and bird migrations. If these things followed a set course, they reasoned,

If a groundhog sees its shadow on February 2 people predict another six weeks of winter. *Spectrum Stock*

weather should too. For many of these weather pundits, the annual meteorological event of greatest economic and social importance was the severity or mildness of the approaching winter.

Today, most meteorologists scoff at weather superstitions, claiming they lack scientific validity. Long-range proverb meteorology, they say, is simply not all right for all times in all places. They *are* silly or foolish notions; however, these light-hearted fancies are fun to repeat and wonder about. They bring public renown to their creators. That's why there is no shortage of long-range weather prophets and believers.

Nonetheless, many naturalists have long subscribed to the theory that certain animals and insects have instincts and sensitivities that make them trustworthy forecasters. One such example is the woolly bear caterpillar famous for its forecasting prowess, especially after its cameo appearance several years ago on "The Tonight Show." Superstition has it that if the brown stripes are narrower than the black, winter will be cold and blustery. Science, however, says the width of the stripes is determined more by a combination of genetics and environmental

## Some Foreign Winter Weatherlore

- When the hare's coat is thick, the winter will be hard (Germany)
- A heavy snow in November will last till April (France)
- Snowy winter, rainy summer; icy winter, hot summer (Russia)
- White Christmas, green Easter; green Christmas, white Easter (Belgium)
- When the leaves of wheat are narrow and short, there will be much snow (Japan)
- Long icicles foretell a long spring (Russia)
- Mushrooms galore, much snow in store; no mushrooms at all, no snow will fall (Germany)
- Autumn thunder means a mild winter (Norway)
- If cattle and sheep after fruit-laden autumn dig the ground and strain their heads towards the north wind, then expect a very stormy winter (New Zealand)
- When gnats swarm in January, the peasant will have an empty granary (Netherlands)
- Mosquitoes in late fall, a mild winter for all (Russia)

conditions when the insect is growing rather than indicative of any privileged insider information.

There is also a widespread belief that an abnormally thick coat of fur on beavers, dogs, bears, and other animals forewarns a grim winter. An animal may well grow a heavier coat if October is especially cold, but that does not necessarily mean the entire winter will be harsh. Again, it's more likely a matter of diet, health, and day length prior to the onset of winter.

> When squirrels early start to hoard
> Winter will pierce us like a sword.

In a similar vein, there is no connection between the number of cones and mushrooms a hungry squirrel or chipmunk tucks away in the fall and the coming winter's severity. Most just tuck away as much as they can in the best hiding place they can find. A large cache may indicate nothing more than a favorable harvest that year. Nor is there any connection between deep snows and places where harvesting creatures store their booty—high up in trees or deep in the ground.

Farmers often turn to their own barnyard animals as weather soothsayers. Pigs and backyard geese have long served as a farmer's weather vane. Ed Burt relies on the shape of the spleen of a freshly butchered pig, as his father and grandfather did before him.

"A thick bulge at one end that tapers off means winter will be short," he explains. "No bulge means an even winter, and an extra wide bulge along the spleen, expect a very cold winter." A cold winter with a mild spell is indicated if the bulge is irregular, he adds.

Then there's the Thanksgiving turkey. According to legend, if half the breast of a cooked Thanksgiving turkey is brown and half white, winter will first be cold, but will warm up in January or February. Other turkey lore suggests that a thin breastbone means a mild winter, a thick bone a severe winter, a light bone more snow but a mild winter, and a dark bone signals colder temperatures are on the way. This last claim is not without merit, for the dark color means that the bird has absorbed a lot of oil, a natural protection against the cold.

Hunters and fishers notice natural omens too. Anglers will tell you to prepare for a bad winter if trout feed voraciously in the fall to gain weight. Hunters pay attention to early southbound geese, the thickness and strength of waterfowl feathers, and how far down a partridge's leg its feathers extend, all to determine winter's clout.

Plants and trees are sensitive to weather and its changes, but they can only convey something about past and present circumstances of growing, nothing about the future.

> Onion skins very thin
> Mild winter's coming in.
> Onion skins thick and tough
> Coming winter cold and rough.

The fact that an onion has many thick skins does not mean that it will be a hard winter, but merely that certain conditions of heat and moisture occurred while it was growing.

## Some Plant Winter Weatherlore

- Extremely tall weeds in the summer are supposed to mean we'll be doing a lot of shoveling
- The thickness of corn husks in the fall is said to be proportional to the degree of winter cold, and a double husk is said to forebode a winter of exceptional severity. Tall corn stalks are proportional to the depth of the coming winter snowfall
- When leaves fall early, fall and winter will be mild; when leaves fall late, winter will be severe. If trees hang on to their leaves, the coming winter will be cold

Still, folklorists see portents in the supply of fruit on shrubs and trees in the fall. They believe nature provides an extra store of berries to feed the birds when deep snows are in the offing. If a bad winter lies ahead, birds will leave the berries on the bushes till the snow comes; if a mild one is coming, they will gobble them up. The reality, of course, is that favorable flowering weather prevailed during pollination and so ensured an ample berry crop.

Another method of seasonal weather divining involves the principles of persistence and compensation, which are reflected in folk proverbs. In the what-you-get category are:

- A foggy autumn will be followed by a soggy winter
- As the weather in October, so will be the next March
- Fall thunder means a mild winter

In the nature-seeks-a-balance category:

- Rainy summer, snowy winter
- Hot summer, icy winter
- A month that comes in bad goes out bad
- When there is a spring in the winter, or a winter in the spring, the year is never good

## Some Animal Winter Weatherlore

- When bees close their hives, a cold winter arrives; if they don't shut the door, a mild winter's in store
- When you see a beaver carrying sticks in its mouth, it will be a hard winter
- Muskrats building houses early indicate a tough winter
- Wild geese moving south: cold weather ahead; moving north: winter is nearly over
- A bad winter is betide, if hair grows thick on a bear's hide
- If yellowjackets start acting clumsy, and crickets sing their last song as the new moon rises, winter will be early

Analyses of lengthy climate records fail to disclose any relationship between summer and winter weather. The only long-term weather sayings with any truth are the cause-and-effect ones, such as:

> A year of snow, a year of plenty.

Or, phrased differently:

> A year of snow, crops will grow.

These are pleasant ways of pointing out that a snowy winter provides enough moisture to assure good crops, and that a good covering of snow insulates crops against killing cold and the cycles of thawing and freezing, especially ruinous to winter grains.

One more maxim that works because it is logical is:

> As the days lengthen, so the cold strengthens.

This simply recognizes that January and February are usually the coldest months, even though the days are lengthening after the December 21 solstice.

Giving nature a run for its money in long-range predictions is that yellow book of weather wisdom, the *Old Farmer's Almanac*. The founder of the Almanac was Robert B. Thomas, who devised the secret formula in 1792 for predicting weather. Popular for generations, this publication purports to know what the weather will be for an entire year, supposedly by using its scientifically enhanced secret formula to blend information on solar activity and climatological data.

"We believe nothing occurs haphazardly," says the *Old Farmer's Almanac* website. "There is a cause-and-effect pattern to all phenomena, including weather. It follows, therefore, that we believe weather is predictable."

The credibility of almanacs springs from people focusing on unusual events, and remembering the successes and forgetting the failures. Furthermore, the weather somewhere on any given day is bound to turn out exactly as called for in at least one almanac. Maybe that's why on the *Old Farmer's Almanac* website (www.almanac.com) you can find anxious brides-to-be asking what the weather will be like on their wedding day, many months away.

## Storm Brewing? The Nose Knows

Lauchie McDougall of Wreckhouse, Newfoundland, had such a keen nose for the weather that he was actually paid to smell approaching storms. McDougall was a part-time cattle farmer and trapper, but it was his job as "gale sniffer extraordinaire" that made him a Newfoundland legend.

Wreckhouse is located on a barren stretch of the Trans-Canada Highway in southwest Newfoundland. It is an area known for its high winds, with gales as strong as 88 mph gaining strength as they sweep down from Table Mountain across the highway and out to sea. Wreckhouse winds are notorious for lifting freight cars off their tracks and blowing over tractor trailers.

Thus the Newfoundland Railway Company and later the Canadian National Railway contracted McDougall to report on the winds for a salary reported to be a maximum of $140 annually. Three or four times a day for 30 years, until his death in 1965, McDougall would sniff out gusts of wind down by the railway tracks and warn officials if he sensed super winds brewing. All nearby trains would then be kept outside the danger area or chained to their tracks until the winds abated. Whatever McDougall said was regarded with great trust.

One railroader recalls, "Many's the train had to stop because Lauchie said so." Once, when officials failed to heed his warning, 22 cars were blown off the tracks.

Following his death, McDougall's wife, Emily, continued to sniff out winds from Table Mountain until she retired in 1973. In 1982, CN erected a plaque in the terminal building in a nearby town, commemorating the work of the McDougalls—who surely had one of the oddest professions anywhere.

**The word *ozone* derives from the Greek for "smell," because of the characteristic odor it gives off when sparked during an electrical discharge. Had a whiff of ozone? Smell that pungent clean odor after a thunderstorm or the "electric" smell of a subway train, and you've smelled ozone gas.**

We have all met people whose sensitive noses are able to smell an approaching storm or rain coming. Although he's no Lauchie McDougall, Keith Fraser, a geographer, says that he can always smell rain outside his home before the first drops reach the ground. With an easterly wind, he smells the sour odors from a nearby pulp mill.

Animals sometimes seem to have far greater sensitivity to approaching weather than humans do. Farmers swear you can count on a rainstorm coming if cows huddle together or lie down in a pasture, or if horses stand with their tails to the wind or roll over. An old saying says, "When sheep collect and huddle, tomorrow will become a puddle." And pigs are said to be especially sensitive, moving sticks and straw before any rain, and squealing at the first signs of a change in the weather. Virgil, the Roman poet, wrote of pigs "tossing their snouts" when a storm was near.

Long before weather forecasters talked of isobars, shepherds, sailors, and settlers developed theories about the smell of rain. It was said that flowers smell sweeter just before a rainstorm and tobacco pipes and manure piles stronger. One old-time rhyme says, "When the ditch and pond offend the nose, then look out for rain and stormy blows."

And sure enough, the decay of organic debris in stagnant ponds, drains, gutters, and ditches produces methane and other gases that accumulate in pockets and bubbles under the mud. These gases and their odors are suppressed under high pressure, but when low-pressure systems (usually associated with stormy weather) approach, the bubbles of putrid gas expand, rise to the surface, and break loose in sufficient quantities to give the local atmosphere an offensive smell. The same lowering of pressure may be marked by the rising of water in wells or toilet bowls, or by more foam on rivers.

**"I believe the behavior of scent depends on two things—the condition of the ground and the temperature of the air—both of which I apprehend would be moist, without being wet. When both are in this condition, the scent is then perfect and vice versa, when the ground is hard and the air dry, there seldom will be any scent." —Peter Beckford in** ***Thoughts on Hunting*** **(18th century)**

Aristotle noted more than 2,000 years ago that shepherds believed that rain became sweetly scented from its passage through the heavens, and that even rainbows had an aroma. Yet we cannot really smell rain because water has no odor. Some meteorologists attribute the smell of thunderstorms to the pungency of ozone; lightning can split apart air molecules, which recombine to form ozone, leaving a sharp but short-lived odor. But several possibilities exist as to why human beings can smell rain coming.

It could be all in the nose. Higher humidity, such as is usually associated with rain, sharpens our sense of smell, and everything smells stronger when it gets damp. As the moisture content of the air increases just prior to rain, the aromatic molecules of many substances become covered with a layer of water molecules. And these larger, hydrated molecules cling more easily to the mucous surfaces inside the nose.

The hydration of aromatic molecules may explain why animals and insects communicate through scents. Masters of fox hunts claim that hunting scents depend mostly on the water content of the ground and air, both of which should be moist without being wet. When the ground is hard and the air is dry, there is seldom any trail for the hounds to follow, but rising humidity improves the line of scent.

**Near Lucknow, India, clay disks are placed outside during the humid months of May and June. The smell of the "rain"—actually the soil—is absorbed, distilled, and bottled. Then it's sold as *matti ka attar*—perfume of the Earth.**

Many botanists believe that the smell associated with rain on land originates with volatile substances given off by vegetation—terpenes from pine forests, for instance, or creosote from desert bushes and various organic smells from meadows. (This, surely, is what sometimes enables mariners to follow their noses through fog—the difference between the heavy salt smell of a sea breeze and a sweet garden smell from a land breeze.) Scientists estimate that plants and trees exude some 450 million tons of plant volatiles, predominantly terpenes, into the atmosphere each year, mainly during the summer months. Apparently, when relative humidity is high or the plant is wet, plant stomata enlarge, and this leads to an increase in the escape of these substances into the atmosphere.

In the 1960s, Australian chemists found that certain types of clays exude a strong "rain" smell when relative humidity exceeds 80 percent. They found that the smell comes from a yellowish oil trapped in rocks and soil, called petrichor—derived from the Greek words *petros*, "stone," and *khor*, the ethereal fluid that flowed like blood in the veins of the gods. Petrichor is commonly observed as the pleasant, refreshing, earthy odor that frequently accompanies the first rains after a warm, dry period.

**Test for yourself the effect of humidity on odor. After taking a shower, compare how strong a perfume smells in the misty bathroom with how it smelled before turning on the water.**

Petrichor comes from atmospheric haze, which contains the terpenes, creosotes, and other volatile compounds that emanate from plants. In the air, these substances undergo oxidation and nitration before being absorbed in the soil or trapped in rocks. Absorption is usually highest when relative humidity is at its lowest. But when the relative humidity climbs above 80 percent, the moisture in the air begins to fill the pores of the rock and the spaces in dry clays with water, displacing the odorous and volatile compounds that then enter the air. Because the earthy scent is so often followed by rain, we learn to associate the two.

Smelling rain, seeing a distant thundercloud, and feeling dampness in the air may not by themselves mean rain, but when combined with hearing a weather forecast, they are almost a sure sign of rain within an hour.

## Be Your Own Weatherperson

During the first half of this century, professional rainmakers (called professors) would travel across the Great Plains of North America selling rainmaking in drought-stricken areas. They often unveiled a weather machine called a *universcope* that spouted colored smoke upward into the sky. Of course, the rain professors would read the sky and then stoke the machines only when there was a good chance of rain. These early cloud seeders knew that for their weather prophecies to have any chance of success, they had to be able to read the sky, assess the current weather, and guess how it would behave over the next couple of hours, which for their business was long enough. A professor with a good weather eye and good luck would soon develop a good reputation.

The former Boston television news reporter Jack Borden is a kind of modern-day rainmaker, peddling clouds instead of rain. Borden is founder of For Spacious Skies, a non-profit educational organization dedicated to increasing sky awareness in people's lives.

**The highest clouds in the atmosphere are called *noctilucent*, or luminous night, clouds. These thin, streaky clouds are very rare, found at heights above 50 miles, where the temperature is consistently between -103°F and -130°F and where water vapor is non-existent. Generally, noctilucent clouds can only be seen at twilight when sunlight reflects off ice crystals or ice-covered dust that lies at the outer limits of the Earth's atmosphere, and usually only in late summer.**

More than 20 years ago, on one delightful June day in Boston, with the sky a rich blue and speckled with fluffy white clouds, Jack Borden conducted a series of person-on-the-street interviews for the evening news, requesting that interviewees "without looking up, describe the sky." To his astonishment, none could.

**Studies show that crickets chirp faster when it's warm than when it's cold. If you count the number of cricket chirps in 14 seconds, then add 40, nine times out of 10, you'll have the temperature within one degree Fahrenheit. It's also said that rattlesnakes can indicate the temperature by their rattle. The frequency of a snake's rattle is zero at 32°F, and about 100 at 98°F. That is, the frequency rises by 1.5 rattles per degree Fahrenheit. These techniques, however, don't work well in January!**

That experience changed Borden into a self-described "Johnny Appleseed of the Sky." Borden contends that except for spectacular events like sunsets, halos, and lightning storms, most people never see the sky. "We breathe the sky an average of sixteen times a minute, but most of us are unaware of its beauty, majesty, power, and fragility." Across America, he has brought the sky into classrooms and Sunday schools, nursing homes, prisons, mental hospitals, and rehab centers, introducing people to the beauty above and the world around by simply encouraging them to look skyward.

Weather should be important to us, beyond the obvious reasons of knowing what to wear or when to mow the lawn. Being weatherwise can help us to protect our property from damage and prevent us from becoming weather statistics. Further, because weather knows no borders, it teaches the lesson of how interconnected we are with one another. Weather is also a chance for young people to develop sensitivity to what's really going on around them.

Watching the sky and its contents can be inspirational, comforting, and soothing. Guessing how it will change over the next little while can be great fun. Furthermore, it requires no special equipment or transportation. It's free! And it all begins by looking up.

A century or more ago, skying was a daily activity. Everyone had to be a weatherperson. Outdoors people—shipmates, trappers, and farmhands—people whose lives and livelihoods depended on coming storms, the thickness of the fog, the breakup of river and lake ice, and the maturing of crops—were seasoned weather forecasters, judging for themselves present and approaching weather.

It used to be that reading the sky and watching the clouds were valuable skills handed down through generations. Now those skills are all but extinct. Regrettably, nowadays most people take weather for granted. Few of us take time to lie down on the grass to see shapes in clouds or watch the ever-changing color of the sky. It's so much a backdrop for daily events that we habituate ourselves to it—like Musak, Borden says. Instead of going outside and gazing at the western sky, sticking a wet finger into the wind, sniffing for rain, or feeling the dampness in the air, we are more likely to turn on The Weather Channel, log on to the Internet, or simply ask someone else.

**Clouds travel thousands of miles horizontally, but no more than about 11 miles vertically. They cover half the world's sky at any one time, but account for a tiny amount of the Earth's moisture; squeeze them all out and you'll get a depth of just a little over an inch to cover the planet.**

Bob Dylan was right when he said, "You don't need to be a weatherman to know which way the wind blows." Through careful observation and study, people can often make their own forecasts for the next few hours.

All prophecies, though, must begin with skywatching. By noting the shape and texture of clouds, the color of the sky, the speed and direction of the wind, and temperature and moisture changes, you can do a reasonable job of predicting coming storms, morning frost or dew, and evening calms. Instead of relying on television broadcasters alone, step outside and expose all your senses to the atmosphere.

Over time, it is possible to develop a few basic forecasting rules that you can use to adapt the official, regional forecast to suit your local situation or circumstances. Here's a handy list of practical weather signs that can hint at the weather ahead. They are generally applicable to many locations in North America, year-round, although the amateur forecaster would want to adapt the guidelines in accordance with local controls such as the moderation of water bodies, changes in terrain and landscape, and urban/rural effects.

Look for cloudy, unsettled weather when:

- The barometer falls steadily
- The wind blows strongly in the early morning
- The temperature at night is higher than usual (sky is cloudy)
- The temperature is far above or below normal for the time of year
- Clouds rapidly move in various directions at different levels
- High, thin, wispy clouds (cirrus) increase in amount, thicken and lower, sometimes producing a ring or halo around the sun or moon
- Clouds darken on a summer afternoon
- High- and/or middle-level clouds darken and move from the south and southwest
- The sunrise is red

Look for steady rain or snow when:

- The barometer falls steadily (if the pressure falls slowly, rain or snow will come within a day; if it falls rapidly, expect precipitation any minute)
- Winds blow from the southeast to the northeast and north
- Clouds are low and uniformly flat and grey
- Leaves show their undersides, as a strong south wind in advance of the rain flips the leaves over
- There is a ring around the sun or moon

Look for more pleasant weather when:

- The barometer is steady or rising slowly
- A gentle breeze blows steadily from the west to the northwest
- Winds swing from the south to the southwest, or from the east or northeast to the northwest
- The amount of cloud cover and the number of clouds decrease in the late afternoon

- The cloud base rises and humidity decreases
- The evening sky is clear and you can look directly at the setting sun, which resembles a ball of fire
- Morning fog breaks within two hours of sunrise
- The night before heavy dew or frost occurs
- The moon shines brightly and the wind is light
- There is a bright blue sky with high, thin wisps of cloud

Look for clearing skies when:

- The barometer rises
- The wind shifts to any westerly direction (especially from the east through south to the west)
- The temperature falls rapidly, especially in the afternoon
- Increasing breaks occur in the overcast
- Clouds become lumpy
- Dark clouds become lighter and steadily rise in altitude
- Fog lifts before noon
- Frost or dew is on the grass

Look for showers (thundershowers) when:

- The barometer falls
- Winds blow from the south or southeast
- The morning temperature is unusually high, air is moist and sticky, and you see cumulus clouds building (rain within six hours)
- Dark, threatening thunderclouds develop in a westerly wind
- Thick, fluffy clouds (cumulus) develop rapidly upward during early afternoon
- You hear loud static on your AM radio (thunderstorms within the hour)

Look for heavy snow when:

- The air temperature is between 14°F and 30°F
- The barometer falls rapidly
- Winds blow from the east or northeast
- A storm lies to the south and east of you

Look for temperatures to rise when:
- The wind shifts from the north or west to the south
- The nighttime sky is overcast with a moderate southerly wind
- The sky is clear all day
- The barometer falls steadily (in winter)

Look for temperatures to fall when:
- The barometer rises steadily (in winter)
- The wind shifts into the north or northwest from the south
- The wind is light and the sky is clear at night
- Skies are clearing, especially in the winter
- Snow flurries occur with a west or north wind

Look for fog when:
- Warm winds are blowing humid air across a much colder surface (either land or sea)
- The sky is clear, the winds are light, and the air is humid the night before
- Warm rain is falling ahead of the warm air
- Water temperatures are warm and the air is much colder

## Keeping Your Own Weather Records

Weather observers make a daily record of weather conditions in their locales at the same time each day. You can learn a great deal about weather and confirm the day's forecast by keeping a weather diary.

At regular times each day, fill in observations of weather, cloud type, and other sky observations. Note the wind direction and estimate its speed using the Beaufort wind scale (an explanation of which appears on page 176). Include a mention of how the weather changed from day to day or from night to day. Make space for personal observations of interesting weather events such as hail, fog, heavy snow, thunderstorms, strong winds, ice pellets, and heavy frost. Instrument recordings of temperature and precipitation at nearby stations can be taken from the local newspaper or television reports.

Make note of the occurrence of other non-regular events such as snow shoveling and grass mowing and annually recurring events in nature such as the dates of the flowering of lilacs, first sightings of robins or other migrating birds, the emergence of insects, or the impact of the first frost on your garden or flowerbeds.

When you supplement the observations with a regular look at the weather map published in most daily newspapers or seen on local television weathercasts, you'll get a closer sense of how weather works.

By watching the sky, noting down your observations, and studying daily weather maps, you'll soon gain additional insight into how the weather develops locally.

Here are three examples:

- When the overnight temperature is forecast to fall close to freezing, local conditions determine whether a killing frost will develop.

Expect morning frost when the sky the night before is clear, the winds are light, and the air is humid. The killing frost may not materialize if a protective mantle of clouds forms or a stirring breeze starts blowing during the night.

- When dew, frost, or fog appears on the grass in the early morning, there is a good chance the day will be fair and bright.

On clear, cool, calm nights, ground moisture in the form of frost, dew, or fog may form more readily because clouds are not present to interfere with ground cooling. Calm, clear nights are typical of high-pressure weather conditions, and fine weather is likely to continue for at least the next day. Cloud cover keeps the earth from losing heat; therefore, the temperature near the earth's surface cannot cool off enough to condense the moisture in the air and form dew or frost.

- When you can smell odors from a ditch or pond more readily, rain and strong winds are coming.

The air surrounding a center of low pressure, such as a storm, is forced upward with decreasing pressure. This phenomenon is analogous to taking the lid off a coffee pot, where odors held in place by sinking air around high pressure are suddenly released. Accordingly, odors become stronger, whether from a bed of roses or from a pond, particularly when aided by higher humidity.

## Weatherlore—Fact or Fiction?

Long before meteorologists discovered cold fronts and jet streams, people relied on nature to foretell the next day's weather. Farmers, mariners, and hunters showed a keen sense of observation and quickly connected changes in the environment with the rhythms or patterns of weather. They recalled what they saw in the form of short sayings, often embodied in rhyme for ease of remembering.

Most weather folklore is a kind of whimsical silliness, imaginative and often contradictory, far-fetched and unfounded, useless and superstitious and quite harmless, though loads of fun. Many meteorologists scoff at weather superstitions. However, many folk sayings have stood the test of scientific scrutiny, even though they were developed without instruments or knowledge about the causes of weather. Those that have a chance of success are the ones that prophesy daily weather changes. The most useful weatherlore relates the coming weather to one weather sign, such as the character and movement of clouds, the color and appearance of the sky, or the direction and strength of the wind, but weather is so complex that relying on one element alone is by no means a sure thing.

**A sudden rise in wind speed may foretell a change in the weather. Likewise, a quick shift in wind direction indicates a turn in the weather. On the whole, winds from the east and south are foul-weather winds; northerly and westerly winds are fair-weather winds. That's why common wisdom says that "it's best to do business when the wind is in the northwest."**

Many people have always found separating wisdom from superstition in weather folklore great fun! Here are some favorite weather sayings; some are about as silly as it gets, and others should be believed:

Dew on the grass, rain won't come to pass.

Evening red and morning gray, two sure signs of one fine day.

The sudden storm lasts not three hours.

The sharper the blast, the sooner 'tis past.

The higher the clouds, the better the weather.

Cold is the night when the stars shine bright.

Sound traveling far and wide a stormy day betide.

Rain long foretold, long last,
Short notice, soon will pass.

If bees stay at home, rain will soon come.
If they fly away, fine will be the day.

A rainbow afternoon,
Good weather coming soon.

Catchy drawer and sticky door,
Coming rain will pour and pour.

The winds of the daytime wrestle and fight,
Longer and stronger than those of the night.

When down the chimney falls the soot,
Mud will soon be underfoot.

When the chairs squeak, it's of rain they speak.

The squeak of the snow will the temperature show.

## The Importance of Clouds

Today's clouds announce tomorrow's weather. By becoming familiar with various forms of cloud and their movements, you'll also be able to make good guesses about approaching weather.

Generally speaking, clouds moving from the south herald precipitation, and those from the north signify clear weather. Also useful to know is that the higher the clouds, the finer the weather. This stems from the fact that high, thin cirrus clouds are not generally associated with precipitation. However, they are often precursors, and depending on the situation, precipitation can quickly follow.

Some additional weather tips from clouds:

- The more cloud types present, the greater the chance of rain or snow
- Clouds during the day can often lower high temperatures; at night they will often trap heat, making for warmer minimum temperatures
- Higher cloud layers move aloft with prevailing winds that carry weather systems; lower clouds reflect local influences

The three main types of clouds to recognize are high level, middle level, and low level. The characteristics of the 10 major cloud types include:

*High Clouds—(above 20,000 to 26,000 feet)*

1. *Cirrus.* Thin, wispy, small white clouds that often occur as feathery filaments or long streamers resembling jet contrails stretching across the sky. Often their ends are swept by strong winds, giving them the look of a horse's tail. Cirrus clouds are fair-weather clouds. If they thicken or spread out into sheets across most of the sky, expect rain or snow within 48 hours. If they do not thicken, fair weather usually continues.
2. *Cirrostratus.* A milky-white, uniform veil of thin, transparent cloud, comprising ice crystals. Sometimes accompanied by a halo around the sun or moon, which usually means wet weather within 12 hours.

3. *Cirrocumulus.* Thin bands of either continuous or patchy small clouds, white or pale gray in color. A ripple or rib pattern gives them the look of fish scales, referred to as a mackerel sky. They signal a change to heavier cloud cover usually within a day. When they are followed by lower and thicker clouds, look for rain and warmer temperatures.

***Middle Clouds—(between 8,200 and 20,000 feet)***

4. *Altocumulus.* Either patchy or continuous middle cumulus clouds with a dappled or rippled appearance, often with a flat bottom. They are considered a thicker and lower version of cirrocumulus. Altocumulus clouds are reliable indicators of changeable weather and an impending storm.
5. *Altostratus.* A pale gray, uniform layer of cloud that is much thicker and lower than cirrostratus. Although they are too thick and low for halos to be formed, the sun can still be seen weakly through the overcast. Usually a reliable sign of precipitation within a few hours.
6. *Stratocumulus.* Low layers of gray or whitish clouds with occasional dark patches that have a well-defined, rounded, or undulating appearance. Stratocumulus may have a few breaks, but usually cloud cover extends for hundreds of miles. They signal changing weather, often preceding a cold front and possible thunderstorm.

***Low Clouds—(below 8,200 feet)***

7. *Stratus.* A uniformly gray, opaque blanket of cloud that may be continuous or patchy, often producing light drizzle or flurries. Its low base often obscures hilltops and tall buildings. Without much texture, it looks like high drifting fog, making for a dull, gray day. At the earth's surface, stratus is simply fog. Fair weather usually follows the disappearance of stratus.
8. *Nimbostratus.* A dark, thick, monotonous deck of low cloud, providing continuous rain or snow. Usually covers the entire sky, hiding the sun. May mean days of steady rain.
9. *Cumulus.* Puffy, white, heaped or piled clouds that often form

## The Beaufort Wind Scale

Sir Francis Beaufort of the British Royal Navy devised the wind-wave scale in 1805. It originally referred to the amount of sail a full-rigged ship could carry in specific wind conditions. In light wind, just one sail would be taken in; in a heavy storm the number would be 11, therefore, Beaufort force 11.

The Beaufort scale has been modified several times. Basically, the idea is to estimate wind speed by watching the effects of wind on such things as flags, trees, smoke, water surface, and even people.

| Beaufort wind force (mph) | Wind speed | Wind type | Descriptive effects |
|---|---|---|---|
| 0 | below 1 | Calm | Smoke rises vertically |
| 1 | 1–3 | Light air | Smoke drifts slowly |
| 2 | 4–7 | Slight breeze | Leaves rustle; wind vanes move; wind felt on faces |
| 3 | 8–12 | Gentle breeze | Leaves and twigs in constant motion; wind extends light flag |
| 4 | 13–18 | Moderate | Small branches move; breeze raises dust and loose paper |
| 5 | 19–24 | Fresh breeze | Small trees sway |
| 6 | 25–31 | Strong breeze | Large branches in continuous motion; utility wires whistle |
| 7 | 32–38 | Near gale | Whole trees in motion; wind affects walking |
| 8 | 39–46 | Fresh gale | Twigs and small branches break off trees |
| 9 | 47–54 | Strong gale | Branches break; shingles blow from roofs |
| 10 | 55–63 | Storm | Trees snap and uproot; some damage to buildings |
| 11 | 64–73 | Violent storm | Property damage widespread |
| 12 | 74 + | Hurricane | Severe and extensive damage |

by day, disappear by night, and re-form the next day. The flat, well-defined base begins around 2,000 feet up in winter and 4,000 feet in summer. Cumulus clouds are associated with fair weather, blue sky, and no precipitation.

### *Vertical Cloud*

10. *Cumulonimbus*. Mammoth cumulus clouds with a dark base and a smooth anvil-shaped top. Called the kings of the sky, they are the biggest of all clouds, often towering in excess of 7.5 miles. CBs are storm clouds associated with severe thunderstorms, heavy rain, and sometimes hail or tornadoes.

# Weather—More than Tomorrow's Forecast

# When in Doubt, Blame It on the Weather

What a convenient excuse the weather is—sometimes legitimately—for events gone awry, for our ailments, or for bailing us out of commitments. "It's too hot," "it's too cold," or "we can't do that because a storm's coming" are about the handiest excuses we'll ever have. It's no longer "the devil made me do it" or "there's something in the water." Rather, we blame El Niño and ozone holes.

Why is weather our favorite excuse? Because no other factor, except perhaps health, looms larger in our daily lives and so directly affects our actions. Weather affects what we eat, how we feel, and how we behave. Weather also plays a large part in influencing market supply and demand: it can create inflation, financial crisis, and social unrest. It is of enormous economic consequence to activities like farming, recreation, energy production, and transportation. It provides 365 different excuses each year for what goes wrong or can't be explained.

**The most imaginative use of the weather excuse has to go to the chairman of a South African company manufacturing cat litter. Explaining the lower profits from his company's business, the CEO blamed "the extremely dry summer, which meant cats spent more of their time outside."**

David Taylor, chief climatologist for Weather Services Corporation, a Massachusetts firm that supplies weather forecasts for businesses around the world, claims that "some companies hire us because they can blame us." That's why weather blaming is as familiar in New York as it is in New Delhi. It is still conversation's first topic, running well ahead of politics, relatives, and sports.

Weather has great appeal as a scapegoat because it is impersonal, random, complex, and uncontrollable. Nothing can be done about it. No guilt—it's nature's fault! Why do airlines so often use weather as a reason for flight delays? Because no one blames the airlines—passengers just curse the weather hundreds of miles away.

Unlike worn excuses such as traffic and family or the dog ate my homework, blaming the weather goes unchallenged. Instead, it invites similar or, better yet, more horrible life experiences from our listeners: black ice, frozen pipes, and baseball-sized hail. Those who demand scapegoats for the weather are left with nations or groups of people. Americans blame Canadians for cold weather and Canadians blame Siberians. For years, any weird, wild, or woolly weather was blamed on the Russians. The end of the Cold War has ended those excuses. Even cold-loving Russians are saying that the winters were better under the Communists. In winter 1992–93, a lengthy mild spell and rare winter thunderstorms brought slush to roads and sidewalks.

**Some of the hottest weather in St. Petersburg, Russia, this century occurred during the summer of 1995. Vladimir Zhirinovsky, the ultra-nationalist leader, blamed the heat on the West. He called on all true patriots to resist this meteorological aggression.**

Of course, there is often good reason to blame the weather. Every day it causes misery, hardship, and misfortune somewhere in the world. According to the United Nations, from 1967 to 1992 weather-related events killed about 3.5 million people and further affected 2.8 billion. Losses from these disasters in 1998 alone amounted to over $90 billion.

Too often, however, weather gets blamed when closer inspection reveals that the problem lies elsewhere. In January 1995, mild weather in central Ontario left usually safe lakes and rivers with huge patches of open water or thin ice that could not support snowmobiles. So weather was said to be the cause of a record number of deaths. But don't the snowmobilers bear some responsibility? Some people exploit weather as a reason to do something unpopular—like raising coffee prices 50 percent when frost kills 10 percent of the Brazilian coffee beans. Others use it as the perfect substitute for unknown causes; they suggest, for instance, that weather caused a rash of robberies in certain neighborhoods or falling revenues at gambling casinos. Weather is the best "educated guess" excuse! And it certainly beats human failure, especially our own. In early 1996, several restaurants blamed a spate of nasty snowstorms that struck just before dinner on several Saturdays for keeping people at home. Those same storms also apparently kept people from dieting. Weight Watchers saw attendance drop 20 percent at their meetings in January and February, normally the prime diet season.

Fascinated by the frequency and scope of weather excuses in our society, the principal author, David Phillips, began several years ago to collect anecdotes from newspapers and magazines. Below is a sampling, ranging from the mundane to the astonishing.

- Athletes are among those most skilled at picking on the weather. The "boys of summer" often credit unseasonably cold Aprils for stiff joints, batting slumps, and poor starts. Even coaches say that the short Canadian playing season leaves too little time for adequate development of baseball players. However, the weather in Canada is no worse than in places like Massachusetts or Michigan, and they produce a lot of ballplayers.
- In 1991, Indy 500 drivers blamed crashes on cold tires due to unseasonably cold weather.
- In 1994, Canadian bobsledders said humidity on their runners resulted in their poor performance at the winter Olympics in Lillehammer.

Unseasonable weather is blamed repeatedly for economic woes. Time and time again, weather is the sole reason for a rise in unemployment, a jump in the consumer price index, more bankruptcies, a decline in housing starts, dwindling profits, earnings disclaimers, rising costs, and a drop in the balance of payments. Has the weather report become the latest economic indicator?

When retailers need a reason for a downturn, weather is their favorite scapegoat, and it may be a good reason when it comes to goods such as bathing suits, skis, and air conditioners. An early snowstorm does wonders for selling winter boots, coats, and studded snow tires. But is it realistic to attribute to the weather all major slowdowns in sales without considering sticker shock, recession, overexpansion, or competition?

In general, industrial, business, and financial managers largely ignore weather when making economic decisions. However, they are quick to use it as an excuse when things go wrong. Breweries blame flat beer sales on cool, wet summer weather; snowblower manufacturers blame near-snowless winters on their losses; and car companies say hot summer weather sapped new-car sales, as did stormy winter weather. According to the Dairy Bureau of Canada, weather is the most significant factor in ice-cream sales. We can easily see that cool wet weather can melt ice-cream sales, but then why does chilly Manitoba have the highest per capita consumption of ice cream in Canada?

**A warm, dry foehn wind, called a chinook in the Rocky Mountain states and a Santa Ana in California, has been linked with an increase in psychiatric disorders and mental-hospital admissions. People's reaction time is slowed down, and they are more irritable and lethargic. However, some people are energized by the wind: higher divorce and suicide rates have been found to coincide with the occurrence of a chinook.**

Sometimes the link between weather and an economic sector is real: for example, the connection between weather and outdoor recreation and tourism is direct and pervasive and it creates both possibilities and limitations. This is why, in 1995, tourist operators prayed for a huge snowstorm to save the snowmobiling season: there wasn't enough snow for trails, and many tourists canceled lodgings because of unseasonably warm weather.

For those in the resource industries of farming, fishing, and energy, the whims of the weather can spell the difference between prosperity and bankruptcy. The winter of 1993–94's cold weather, with an average temperature two to three degrees below normal, meant increased gas shipments and higher profits for gas utilities. Lingering cold winters can shove fuel prices way up because they draw on the gas supply, and like anything in short supply, the gas tends to cost more. In agriculture, the weather cycle and growing cycle are virtually synonymous. Weather is an important factor in a multitude of farm decisions such as what, where, and when to sow, irrigate, cultivate, protect, and sell; and what equipment to buy or use. To the farmer, weather is the nearest and dearest and most feared companion.

In other instances, the connection—and therefore the rap—is not today's weather, but conditions that occurred last season or a year ago. In 1988, Yukon ranchers blamed the weather for an increase in the number of wolf attacks on livestock that summer: rabbits and other traditional wolf food had been killed off by the previous winter's bitter cold. In the summer of 1994, Toronto-area motorists had to endure emergency road repairs, lane restrictions, and detours as crews repaired roadways: the deep freeze of January had been especially hard on roads, resulting in the worst outbreak of potholes in at least 30 years.

**Men and women suffer through the same number of colds each year—2.2 on average—but women feel more miserable, according to a study by pharmaceutical company Smith/Kline Beecham. In a poll of 150 men and 150 women, they found that women more often blame their colds on external causes such as the weather or being around infected people. Men are more likely to blame themselves for not getting enough rest, vitamins, or exercise. For remedies, women choose juice; men opt for a hot toddy.**

Faulting the weather is often the easy way out. Full explanations may be much too complex or involve many factors—or else we just don't know.

People are probably right to blame weather for how they feel. We know, for instance, that weather affects certain diseases such as arthritis and asthma, and is linked to mood disturbances and aggressive behavior. Death rates from heart attacks are much higher during cold months than at any other time. However, it may be unrealistic for people to blame extremes of barometric pressure for making people sleepier or hot, humid weather for the sluggish growth of children.

Too often, weather gets an undeserved bad reputation. Here are some even more surprising examples where weather was claimed to be the main culprit:

- Don't blame the chef for flat pastry: apparently without the right weather puff pastry won't rise

- Colonel Gadhafi claimed weather, not terrorists, was responsible for the jet crash over Lockerbie, Scotland
- Fewer babies are born in spring than at any other time; the explanation given is that men produce less sperm in hot summer weather
- Weather is to blame for spelling errors and opening-night jitters
- People are more easily depressed in humid weather
- Weather was blamed for the poor response to a food-bank drive
- Wet, cool summers mean fewer butterflies and honeybees
- In June 1988, hot, dry weather in Manitoba, Canada, made it a bad year for woodticks—so severe was the tick problem that horse owners had to shoot animals covered with the bloodsucking parasites
- Mild winters have caused the rat population in Britain to increase by 20 percent
- Because of stormy winter weather in 1994–95, more police officers were busy investigating accidents; this meant fewer cars were pulled over in the program to test for drunk driving
- A hot, dry summer in 1988 more than doubled the price of worms at bait shops
- Organizers for a Michael Jackson concert blamed hot weather for Michael's migraine, which led to the cancellation of his stage show in the Far East

Adverse weather thousands of miles away can be blamed for situations close to home. In 1994–95, Hurricane Gordon destroyed winter vegetables and citrus fruits in Florida, and heavy rains did the same in California. The adverse weather meant an immediate doubling and tripling of fresh fruit and vegetable prices in Canada. In December 1989, cold weather across the United States plains and Midwest caused live hog and cattle futures prices to rise. The cold reduces animals' weight gains and causes farmers to slow deliveries of livestock to slaughterhouses.

Thankfully, weather bashing has become pervasive and universally accepted. How else do you think people can get away with not visiting relatives or avoiding doing jobs around the house?

# Being Practical About Weather

Over millions of years, different climates have shaped landforms, laying down carbon deposits, feeding glaciers, sustaining rivers, and producing soils. In addition, there is scarcely one aspect of society and the economy that is not affected, in some way, by climate and climate change and by the day-to-day influences of weather. Foremost, atmospheric changes provide life's essentials: heat, moisture, and light.

The march of diurnal temperatures and moisture and the progression of the seasons influence how we dress, what we eat, how we feel and behave, the cost of heating or cooling our homes, and our vacation plans. By influencing market supply and demand, climate can create inflation, financial crises, and social unrest. It is essential to the production of trees, the growth of crops, the success of fisheries, and the management of water resources. Used to plan and design, weather and climate can make for safer and more comfortable activities and more profitably run operations.

Because it restricts agriculture, fisheries, and forestry to specific geographical areas, climate has influenced human migration and settlement. Through drought and flood, climate can destroy and debilitate life, damage property, and isolate entire communities.

There are many untold examples of good planning through the proper application of meteorology. Just the fact that buildings aren't collapsing daily under climate stresses or that crops everywhere aren't failing is evidence that applying good climate knowledge and information greatly benefits humankind. Climate and weather information is of enormous economic worth in such activities as farming, recreation, energy production, and transportation, providing answers to such questions as: How much should be budgeted for snow removal? Are there sufficient hours of wind blow to sustain a wind-driven generator? Which climates are best for people suffering from asthma or arthritis? Will peaches grow in Juneau? Is there enough snowfall for a successful ski resort? What impact will the effects of global warming have on sea-level rise in Hawaii and traditional native life in the Arctic?

By answering these questions, using decades of weather records

Corn withers in a Montana drought. *Getty Images*

collected by observers, and, more recently, data from satellites, aircraft, radar, and other sources, climatologists serve the needs of many sectors of the economy.

The following sections contain examples of how climate and weather data, information, and services have benefitted those in selected weather-sensitive sectors or activities.

## Agriculture, Forestry, and Fisheries

Weather affects humankind most closely in farming. To be successful, a farmer must minimize risk from climate hazards by avoiding the risk in the first place or by protecting against the hazard when it threatens.

All agricultural areas in the US are subject to drought, frost, wind, heavy precipitation, hail, and flooding, and to climatically influenced diseases and insect infestations. Evaluation of lands where climate is suitable for special crops is an important application of climate knowledge and information. But even when the climate is generally suitable for a particular crop—for example, Florida's climate is great for growing oranges—a weather anomaly like frost can cost dearly. A freeze in central and northern Florida in December 1983 cost the citrus industry about $2 billion. Florida citrus growers had hardly recovered when another freeze in January 1985 struck and did another $1.2 billion worth of damage.

**Guests at a Myrtle Beach, South Carolina, resort can buy insurance against rainy weather, provided they do so 14 days before their vacation begins. The premium against rainy days amounts to $5 for every $100 of coverage.**

Climate and weather influence a host of farming operations, including seeding, irrigating, spraying, cultivating, harvesting, and scheduling labor. Post-harvest concerns about crop storage and transportation and livestock performance are also affected by weather conditions. Few question the importance of accurate and timely weather forecasts to the agriculture and food-growing business. Simple climate data, such as growing-degree days and drying indices, have found wide application in improving many farm-management operations. More complex crop-yield models are also widely used by growers and by agronomists and economists in agencies and various grain cooperatives in order to forecast growing conditions and develop marketing strategies. A recent development is the monitoring of climates and related information abroad in order to capitalize on global trading and marketing opportunities.

Climate is no less a controlling agent in tree production. Foresters seek current weather data and information in protecting against hazards such as drought, excess water, fire, frost, blowdown, air pollution, pests, and diseases. The monitoring of fire-weather parameters—moisture conditions of the soil and ground litter, wind speed and direction, humidity, and thunderstorm activity—are especially crucial in forest-fire prevention and control. And in combating pests and diseases through aerial spraying, synoptic weather conditions, such as the onset and duration of the sea breeze, are crucial in deciding spray strategies.

Longer-term climate information, both historic and projected, are eagerly sought by forest managers because meteorology affects the growth of trees at every stage from planting to harvesting, and influences forest management practices, including site preparation, regeneration, thinning, and fertilizing. Regeneration of forests takes decades, and to ensure future supplies of trees it is important that varieties be selected to optimize climate potential. No better example exists of the need for information on climate change and variability than forestry. After all, trees are likely to be harvested in weather conditions much different from those that prevailed during planting.

The size of fish stocks and their migration patterns are greatly influenced by changes in oceanographic and atmospheric parameters. Climate data could have important economic benefits to fisheries through improved understanding of the onset of ice formation and ice breakup and the distribution and migration of various commercial fish species, and through improved access to fishery grounds. Careful monitoring of currents, temperature, and other oceanographic factors can save millions of dollars because slight changes can have a tremendous impact on the size and migration patterns of fish species in a certain area. A capacity to predict monthly and seasonal variations would greatly enhance the tactical planning of fishing-fleet operations and would improve the management of fisheries.

## Water Resources

Precipitation is the source of all the Earth's fresh water, and evaporation is the primary way of moisture input into the atmosphere. North America contains large areas in the Southwest where annual evaporation greatly exceeds precipitation, producing permanent deserts. Certain regions and large cities—like Phoenix and Los Angeles—have such high water demand that climate variations are of concern to those involved in securing water supplies.

Climate is a principal factor in determining the water-resources potential of a region. Satisfying the demand for water, and in amounts that avoid flood or drought hazards, requires credible climate data and weather forecasts. On the Plains and in the South, farmers anxious about water for irrigation and municipal officials worried about flooding regularly seek information about the winter's snowpack and spring rains. After all, both drought and floods can lead to incredible economic losses. A heat wave in the US South in the summer of 1998 killed both crops and livestock and caused over $6 billion in damages. One of the worst disasters in US history was the drought of 1988, which caused an estimated $40 billion in losses across the Plains and the East. The Great Flood of 1993 cost the Midwest over $20 billion. Heavy spring snowmelt in the Northern Plains caused significant flooding in North Dakota, South Dakota, and Minnesota in the spring of 1997 and left over $4 billion in damages.

Engineers on the Great Lakes make daily assessments of the basin's water balance in order to predict lake-level changes and regulate flows. More precise definition of climate parameters can improve water supply and demand forecasts and in turn ensure enormous savings in power generation and in scheduling drafts of ships on the rivers. Hydroelectricity is big business throughout North America, and electric utilities count on ample winter precipitation to fill reservoirs. Credible climate data can indicate abnormally high water storage in the preceding season or prompt relevant operating strategies to avoid potential shortfalls. In moist regions with a reliable streamflow, engineers develop schemes for diverting surpluses to water-starved regions.

## Energy Consumption

The US consumes more energy per capita than any other country in the world, except Canada. This thirst for energy is caused largely by vast distances, its resource-based economy, and its climate extremes. Thirty-six percent of US households—people who live in the colder areas of the country—spend an average of more than $500 per year on energy (electricity, gas, or oil) for space-heating their homes. In the warmer parts of the country, like the South, average heating costs are about half of what they are in Northern regions, but Southern households spend an average of $275 per year on energy for air conditioning. And households in the moderate zone—almost a quarter of the population—spend an average of $648 per year in combined heat and air conditioning energy costs. In all areas except the warmest, households use more energy to moderate the temperature in their homes than for all other household energy uses combined (water heating, refrigerating, and appliances). In the 1970s and 1980s, severe winters and uncertainty about foreign oil supplies focused attention on the US's vulnerability to climate change and variability.

Further cooling or increased snow would cost the US economy

dearly; on the other hand, a warming of 4°F to 5°F might decrease winter heating needs significantly. The safe and economic exploitation of energy reserves demands good climate information. A large petroleum exploration company drilling off the Labrador coast in 1974–76 found that only one or two days were lost each season by having to move their drill ship off-site to avoid icebergs (despite using other ships to tow potentially dangerous icebergs away from the well site). Had the climate information relating to iceberg size, frequency, and movement not been incorporated into initial design work, a conventional moored drill ship might have been used and could have been completely lost, due to its inability to move off-site quickly enough. A huge capital loss would have resulted, not to mention the possible loss of human life and the potential environmental impact of an oil spill.

**The John Hancock Mutual Life Insurance Company headquarters in Boston lost a third of its insulating, double-glass windowpanes before occupancy. Poor climate design was the principal reason for failure. Every single window was replaced with heat-treated glass, at a cost of $10 million.**

Furthermore, climate-related applications are not restricted to energy exploration and development: electric-power transmission lines are sensitive to icing, lightning, high winds, and tornadoes; and construction and operation of pipelines in permafrost depend on changes in temperature and soil moisture.

Another issue is how much of the US's energy needs can be met by renewable sources. Americans spent about $1,975 per person on energy consumption in 1991. Passive solar heating could reduce energy bills. Inexhaustible, pollution-free wind energy, once very popular, could again supply much of the energy needs of small, remote communities. In achieving energy conservation, climate information is useful in planning potential sites, designing the materials, and operating the facilities.

## Building and Construction

In most countries, the building and construction sector is the largest single industry. From the viewpoint of safety and economy, a structure or facility must be able to withstand meteorological forces and loadings over its lifetime and take steady wear and tear during both normal times and extremely unusual events. Overdesign is wasteful and costly; under-design can be hazardous and result in discomfort.

Architects, engineers, builders, and planners are often guilty of placing structures suitable for one climate in a very different one—for example, California-style homes in New Jersey. Climate information is useful in determining the orientation of buildings so that there is minimum energy loss and reduced snow drifting, and in designing buildings that have comfortable, healthful, safe, and economical indoor climates. At a large hardware distribution center near the international airport in Toronto, loading docks were designed to keep them clear of snow, thus cutting snow-removal bills and minimizing disruption times. In hurricane-prone southern Florida, disaster management officials and insurance companies encourage the use of storm shutters, which can be closed to protect windows from lashing hurricane winds.

**The climate has economic significance. In a typical winter, Buffalo, New York, may spend millions of dollars on snow removal, while Miami will spend nothing. But air conditioning costs will be much higher in Florida. For farmers and ski-lodge operators, the whims of the weather can spell the difference between prosperity and bankruptcy. Because of the extremes of the US climate, Americans are among the world's largest energy users.**

Climate data are used in support of national building codes to obviate structural failures due to climate stresses and to ensure adequate heating and ventilation. A reduction in the amount of concrete to be poured, in the capacity for backup energy supply, or in the thickness of transmission cables and guide wires can save millions of dollars in individual projects.

Such information can also help in the choice of safe, economical materials that will withstand damage by tornadoes, hurricanes, strong winds, driving rain, or frequent freeze-thaw temperature changes. Home-owners can realize greater comfort and reduced energy costs by considering such questions as where to plant trees in order to provide shelter from winter winds and to keep driveways free of snow; what sort of curtains to choose in order to keep out unwanted summer heat; or which wall of the house should contain a large area of glass to let in winter sun.

## Transportation

Water, land, and air transportation are all influenced by climate variability or extremes. Route planning and scheduling and cargo handling and storage all depend upon sound climate information. In marine areas, sea ice, poor visibility, and storm occurrence are significant problems and affect the design and operation of ships and port facilities. Design must incorporate information about storms—like the one that sank the *Ocean Ranger* oil rig, drowning 84 men on February 15, 1982—in order to defend against such hazards. Metallurgical changes induced by persistent cold led to the failure of a drilling platform off Norway, with the loss of about 100 lives and the capital investment of about $1.25 billion. Better climate data in design could have prevented the collapse.

Freezing and thawing precipitation and snow cover are very important to land transportation by road, rail, or pipeline. Annual snow-removal costs in many cities are in the millions of dollars. Knowledge of the normal distribution of snowfall and its relative frequency of occurrence during the day and over the season is of great assistance to highway-system designers in locating maintenance depots, designing snow-removal equipment, and scheduling work periods for crews. Air transportation has always been concerned with wind, ceiling, visibility, and turbulence, whether for airport siting, timing of favorable takeoffs and landings, or en-route aircraft operations.

## Health and Recreation

One of the most interesting and challenging areas of modern climatology deals with the effects of climate on human health. As we have noted, climate-related health problems include the common cold, hypothermia, hay fever, asthma, frostbite, and migraine complaints. Even more common are subtle climate-induced or climate-intensified sickness, fatigue, pain, seasonal adjustment disorder (SAD), and insomnia, all of which ultimately diminish alertness, learning ability, performance, and productivity.

Climatic factors are also pervasive in the field of recreation and tourism. Many communities, and even entire countries, are dependent on the income from tourism. For developing countries, tourism is a major source of foreign exchange and it provides important income generation and employment opportunities. The economic well-being of vacation areas depends upon the occurrence of expected weather. Travel agencies and resort complexes use climate data to choose suitable sites for business meetings and conferences and the times for peak activity. Downhill skiing is particularly sensitive to the variability of snowfall. The absence of snow generally keeps people from skiing despite snowmaking equipment, and a lack of snow between Christmas and New Year's Day can spell disaster for resort operators. Climate information has been used in the siting and design of trails and the design of equipment and facilities; in the assessment of competing sites for sporting events; and in the scheduling of games and contests. A climatologist advised local resort promoters interested in developing a snowmobile park to avoid the proposed site where only 10 to 20 days a year could be counted on for sufficient snow cover; a second site with favorable snow conditions was recommended, and the resort was built there.

## Industry and Commerce

In spite of the fact that meteorological knowledge and information can profitably be applied to a broad span of industry, commerce, and financial services, most managers in these areas largely ignore weather and climate when making economic decisions. They dismiss weather because they see it operating in some random fashion, and because, they believe, there are too many other more important variables to consider. For example, climatologists can assist insurance companies in establishing the degree of risk for natural hazards (wind, hail, floods) and therefore the premiums to be levied, together with expert advice in settling claims.

Climate information is also used by business and industry to make critical choices concerning delivery and stockpiling of raw materials, production, and marketing, and by stockbrokers and commodity traders in buying and selling.

Most large national retailers are aware of the average dates of significant weather events in each market area and therefore have consumer goods and advertising ready in these localities at these times. Climate-sensitive products such as bathing suits, kites, skis, ice cream, and antifreeze should be marketed near areas where they are most useful and in the appropriate season for successful results. Automobile manufacturers use precipitation data to study windshield-wiper performance on passenger cars, and they change door-handle design to minimize freeze-up for automobiles sold in places with extremely cold winters.

With such an important array of beneficial uses, it is surprising that relatively few people are aware of the enormous potential and economic worth of using weather data and information beyond whether it will rain on the weekend or the coming winter will be long and cold.

Using information about weather and climate wisely will help enhance our management of resources. Used to plan and design, meteorology can make for safer and more comfortable activities and more profitably run operations.

## Frostbite

In the last week of February 1994, a vicious blizzard plunged southeastern Saskatchewan into a deep freeze. At 2 a.m. on February 23, two-year-old Karlee Kosolofski of Rouleau, Saskatchewan, wandered out the front door of her parents' house and was accidentally locked outside in a wind chill that froze her flesh in 30 seconds. Six hours later, her mother found the little girl's apparently lifeless body. Her subsequent recovery, from severe hypothermia and frostbite, is perhaps the most publicized account of cold-weather injury in recent times.

Doctors later estimated Karlee's core body temperature fell to 57.2°F, some 41 degrees below normal. Miraculously, medical staff brought the youngster back to life by artificially warming her blood with a heart and lung machine. Little Karlee made the *Guinness Book of Records* for surviving the lowest recorded body temperature.

Her frostbite injuries were much more serious. Doctors said her legs had the brick-like consistency of freezer meat when she was found. Ten days later, her left leg was amputated below the knee. She also had minor frostbite to her nose, ears, and right elbow.

## Pet-Safety Tips

Pets get frostbite, too, and need extra care on cold, icy days. Here are some pet-safety tips for cold weather:

- Cats are best kept inside, as are most small and short-haired dogs
- On a cold day, apply Vaseline to your pet's paw pads before it goes out. Wash ice, salt, and other chemicals from your pet's pads once it's back inside
- An outdoor dog needs a dry, elevated doghouse with clean, dry bedding and a flap over the opening to keep the wind out
- Check outdoor water bowls often when it's freezing. Do not use metal dishes for food and water because the tongue, nose, and lips can stick to metal surfaces
- Do not allow pets to eat snow on city streets or drink from puddles
- Outdoor dogs need more calories in the winter to produce body heat, so give them more food; indoor dogs and cats may get less exercise in cold months and will need fewer calories
- Antifreeze tastes good to pets but it is a deadly poison
- If it is too cold for you to go outside, it is too cold for your pet

Thousands of North Americans suffer each year from frostbite. Frostbite is the freezing of the skin and underlying body tissue after exposure to temperatures below freezing. It's almost exclusively a human injury. Cold-adapted animals such as wolves, caribou, and polar bears have sufficient blood and heat flow in their extremities to prevent freezing at −94°F. Extremities—fingers, toes, ears, the nose, and chin—are often the first to freeze because they protrude and have relatively poor circulation. Knees, legs, shins, cheeks, and the forehead are also vulnerable. Exposure to extreme cold in itself does not necessarily lead to frostbite. But cold combined with exhaustion, shock, hunger, dehydration, injury, prolonged immobility, or clothing that is inadequate or constricts blood flow can lead to frostbite. Alcohol consumption before exposure is also a factor in many cases.

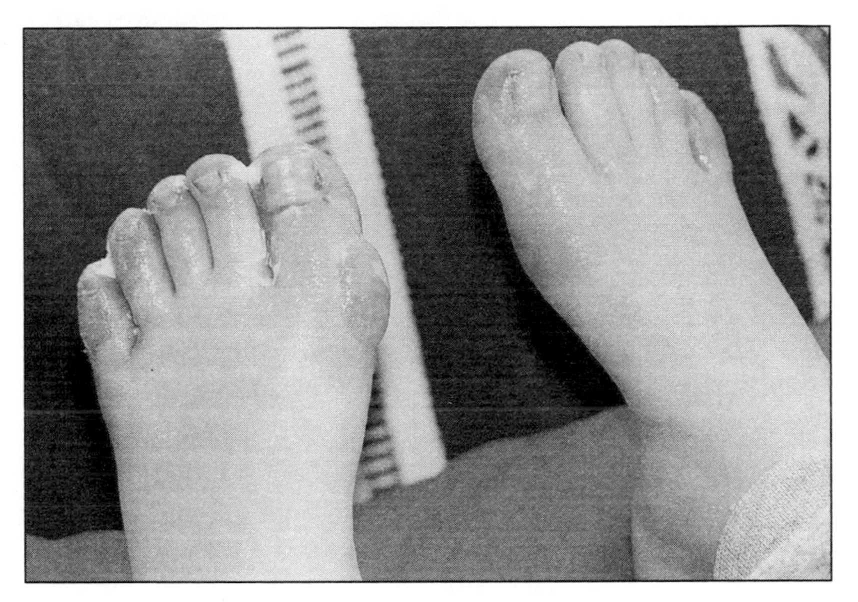

Painfully raw, blistered, and cracked frostbitten feet. *Canapress*

Frostbite has plagued soldiers for centuries, because they live outdoors in all weather, are often wounded or tired, and frequently have insufficient food, water, and warm clothing. The armies of Hannibal and Napoleon lost thousands to the cold. During the Crimean War, French troops suffered more than 5,000 cases of frostbite, and during the Korean War a quarter of all American casualties suffered cold injury, mostly frostbite, trench foot, and hypothermia. Among polar travelers of the last century, frostbite was quite common.

**Elderly people may be threatened by a subtle form of hypothermia—a core body temperature below 95°F—if the indoor room temperature is too cool for them.**

Today, frostbite is much less prevalent than it was even two decades ago. One noted expert on cold weather injury, Dr. Murray Hamlet of the United States Army Research Institute of Environmental Medicine near Boston, suggests a host of reasons for the decline: the public is better informed about cold weather survival, fewer wars are fought at high latitudes, more students are bused to school, and better outdoor clothing, equipment, and central heating are being used. However, for those who work, travel, and play in the cold, frostbite is still a serious risk. Street people are also at high risk as are people stranded by stalled vehicles or accidents. Elderly persons and the very young are particularly vulnerable to frostbite. Seniors are at risk because they are usually less active and unable to move around to maintain their circulation adequately. Children have a large surface-area-to-body-mass ratio and tend to cool fast.

**The Navy manual issued to personnel going to the Arctic and Antarctica says: "Do not touch cold metal with moist, bare hands. If you should inadvertently stick a hand to cold metal, urinate on the metal to warm it and save some inches of skin."**

People in many areas of North America are at frostbite risk with freezing temperatures, moisture, and wind chill. The degree of frostbite depends, naturally, on the severity and duration of cold stress. (Temperatures must be below 31°F before flesh will freeze.) Moisture can contribute to heat loss in two ways. When exposed skin becomes wet, body heat evaporates the moisture, which chills the skin (this is also why we feel a chill while drying off with a towel). Second, moist clothing draws heat away from our bodies, which is why a cold damp day often feels colder than a cold dry one.

Wind chill is a measure of the air's cooling power—the effective temperature of wind plus cold, not simply the air temperature. Our bodies are enveloped in a very thin layer of still air that insulates us against colder outer air. Without any wind, this warmed air stays next to the skin. But that insulating layer can be thinned or blown away by the wind. That brings more cold air into contact with the skin than the body can counteract with its own heat, and we feel a chill. A temperature of 0°F combined with a wind speed of 20 miles per hour feels like −39°F. At that wind-chill temperature, exposed skin can freeze in under a minute.

What happens when you develop frostbite? As your extremities begin to chill, the blood vessels in your skin constrict in order to prevent warm blood from flowing to the surface and losing its heat. Without heat, your skin tissue freezes. The thin blood vessels beneath the frozen tissue also constrict, reducing blood supply. Simultaneously, the body dehydrates as water is withdrawn from the cells by osmosis, thickening the blood. This thickened blood cannot travel through the narrower blood vessels, and the flow is cut off or diverted from the extremity. As the water freezes within the tissues, the ice crystals damage blood vessels, and biochemical changes damage the tissue. Rapid freezing does less harm, since it creates smaller crystals. The fact that Karlee froze solid in 20 minutes may have saved her life.

As with burns, several degrees of frostbite are recognized, although most experts classify injuries as either superficial or deep. Frostnip (first-degree or superficial frostbite) is the mildest form, and seldom leads to permanent injury. At this early stage, the victim is often unaware of the injury, except for some numbness. There may be some local itching or burning and tenderness, and the skin may look pale and waxy.

**You can lose 50 to 70 percent of your body heat through the top of your head because it has so many blood vessels and very little fat. Wear a hat in cold, windy weather. Cowboy hats and baseball caps are okay, but a toque, balaclava, or parka hood are much better.**

Second-degree frostbite freezes tissue beneath the outer layer of skin. The frost-nipped tissue may be cold and hard to the touch. Thawing can bring excruciating pain. Sometimes swelling occurs, and clear, superficial blisters erupt a day or two after exposure; in about a week, the blisters dry, and in a month, a layer of pinkish skin forms.

Often, third- or fourth-degree frostbite involves damage to deeper tissues including nerves, blood vessels, muscle, and bone. With a loss of nerves, there is sometimes an absence of pain. To the touch, the damaged skin feels woody and lifeless. In a few days, ugly dark blood blisters may cover the entire affected area, signaling the death of skin and possibly of deeper tissue. The affected part remains swollen and discolored. Eventually, gangrene sets in and the dead tissue simply falls off.

Once you've had frostbite, you are more likely to get it again, since the blood vessels never regain their full size after recovery. But this increased sensitivity to cold can actually reduce future risks. Sakiasie Sowdlooapik, visitor service officer at the Auyuittuq National Park

Reserve on Baffin Island in Canada's Arctic, confirmed an old Inuit custom of intentionally exposing ears to cause frostnip, in order to acquire the subsequent "early-warning" system for frostbite.

After recovery, frostbite victims often experience persistent pain, excessive sweating, burning, or itching. In the long term, some people report arthritis in affected bones and cancer in scar tissue. Amputation of desiccated tissue may be necessary as a last resort. Unfortunately, amputation is often done too soon, before the tissue has had a chance to recover.

Frostnip can be treated on the spot by blowing hot breath on the affected part; holding nipped fingertips in the armpits; and placing warm hands over the ears or cheeks. Forget that old folk remedy of rubbing snow or applying ice on frostbitten skin: it doesn't work, and just might inflict permanent injury. Someone once said that rubbing ice on frozen skin was like grinding glass into the skin. Nor should you try to warm a frostbitten appendage from a campfire, heat lamp, hot stove, exhaust pipe, hot-water bottle, or other source of dry heat: since your skin cannot feel the heat, you could easily burn yourself.

**In a study of 101 patients suffering frostbite who were admitted to a hospital over the course of 10 winters, researchers found that alcohol consumption was a causative factor in 39 patients and a motor vehicle accident or breakdown in 33 others.**

Deep frostbite is a serious injury that requires immediate medical attention.

Frostbite risk is one reason to respect, but not fear, the cold. Guard against the effects of cold, moisture, and wind by wearing clothing in multiple layers rather than a single heavy outer garment, and especially covering your hands, feet, and head, through which as much as 70 percent of total body heat is lost. Surviving in the cold is a matter of having common sense and being alert, prepared, and properly equipped.

One should always have an eye on the weather. Be aware not only of the air temperature but the wind speed, because strong winds make it feel colder than it actually is and increase the rate of heat loss. In addition, overexertion will produce excessive perspiration that will lead to evaporation and heat loss.

Back at home, young Karlee Kosolofski has been fitted with a

### What to Do If You're Stranded in a Car

- Stay with the car—it's your best shelter
- Make sure that the exhaust pipe is free of snow; otherwise, deadly carbon monoxide fumes can back up into the pipe
- Run the engine for 10 minutes every hour. This should keep you sufficiently warm, while keeping the battery charged and conserving gas
- Keep a window partly open for air when the car is running
- Make your car visible with a flare or warning light. Even a scarf tied to the antenna will help

prosthesis attached below her left knee, and has learned to walk all over again. As for the terrible ordeal, her amazing recovery, and the attention paid her case worldwide, her father, Robert, reports that Karlee would rather talk about school or tease her older sister and brother.

## Profiting from Weather

Here's a hot stock-market tip! Buy when it's raining or overcast and prices are depressed. Sell in the sunshine when traders and investors feel better, buy more, and in doing so bid up the price.

Just like the weather, stock prices on the New York Stock Exchange and other exchanges vary considerably and reflect many things, including the economy, prices and interest rates, the jobless figure, elections, international news, what prominent investors, politicians and bankers say or do not say, and the weather.

Weather? Well, it's not one of the more immediate factors that come to mind in explaining stock-market behavior. Atmospheric pressure and storms are usually lumped in with factors such as sunspots and phases of the moon—somewhere between "market exuberance" and the height of hemlines.

**On February 11, 1994, a severe winter storm dumped heavy snow across the eastern seaboard and forced the New York Stock Exchange, the American Stock Exchange, the NASDAQ, and others to take the highly unusual step of stopping operations. It was the first time weather had halted trading since a hurricane struck New York in 1985.**

Few investors would consider that there is any link between weather events and the rise and fall of the stock market, even though weather can make a big difference in a company's earnings. Because weather is highly related to demand, supply, and production, the financial performance of retailers and utilities, for instance, can be tied closely to weather anomalies. Consequently, entrepreneurs and investors should be concerned with weather, especially before it actually occurs.

Astute market watchers spend considerable time following the highs and lows on the weather news because they realize only too well that weather—especially severe storms and insidious events like droughts and floods—do affect how some companies and key economic sectors perform.

Obviously, some sectors and businesses are more interested in weather reports than others. Breweries, ice-cream parlors, and swimming-pool installers revel in hot weather because it traditionally pushes up sales. Ski-resort operators rejoice at an early heavy snowfall. And investors with stocks in snowblowers and snow shovels, or cold remedies and tissues, get quite excited at an unusual spell of tough winter!

The insurance industry is clearly affected by weather extremes. And though economic losses from natural disasters have tripled over the past three decades, insured losses have increased by a factor of five. A bad year for hailstorms, tornadoes, and freezing rain means insurance companies have to pay more to their policyholders, which has a negative result for shareholders.

But the same storms can drive lumber and building material prices and profits through the ceiling. That is exactly what happened in August 1992 after Hurricane Andrew struck southern Florida. Damage losses, mostly to homes and buildings, cost over $30 billion. But this was a bonanza for the lumber industry in North America, as demand and sales of building materials soared at a time when business is usually slow.

Another industry very affected by weather is agriculture, with its extremely volatile prices. A mere hint of a deep freeze or rumor of coming rains can rattle the markets and send the price of food commodities on a rise and dive as undulating as the jet stream. Traders stand to make or lose millions of dollars, depending on news that it's too wet, too dry, or too cold. Hence, in 1996, the simple beginning of the frost season in Brazil was enough to quadruple the price of coffee beans in New York.

**A finance professor at the University of Massachusetts and former stockbroker analyzed 28 years of weather data and compared the data with the daily indexes on the New York and American Stock Exchanges. On sunny days, he found the indexes were up more than 57 percent of the time, whereas under cloudy skies the indexes were up less than 50 percent of the time.**

At the beginning of the major North American drought in 1988, a meteorologist created quite a stir when he went outside for a cup of coffee. From his office at Smith Barney Inc. in Chicago, Jon Davis inadvertently moved markets when he donned his trench coat. Davis told *The Wall Street Journal*, "I had to walk past the traders to get out, and everyone figured this was some kind of secret signal, that I thought the weather would break and start raining and maybe they should sell the market. In fact, it was a cool day in May and I'd put the coat on because it was chilly."

Those investing and trading in oil and gas also hang on every weather word. When winters are cold, people use more natural gas to warm their houses and businesses. The more natural gas people use, the more profit for the gas company. For two years after the eruption of Mount Pinatubo in the Philippines in June 1991, global temperatures declined by more than 1°F, or by as much as they had risen in a century of warming, enough to produce colder winters in parts of North America and Europe. That single eruption on the other side of the world did more to quadruple the price of natural gas—and earn billions of dollars for the industry—than the Gulf War or any

United Nations embargo. Likewise, late in 1996, the Toronto Stock Exchange's oil and gas index rose by 20 percent, largely because of the winter outlook, which called for a colder three months ahead than normal—a prospect that would likely boost oil and natural gas demand and increase producers' profits.

**Because weather only repeats itself from year to year about 35 percent of the time, retailers who predict sales based on the previous year's weather will be wrong two seasons out of three.**

The commodities market, however, is perhaps the best example of weather sensitivity in the business world. Commodity traders deal in buying and selling contracts to deliver such commodities as soybeans, frozen orange juice, gold, tea, hogs, or natural gas. The goods are never seen by those who buy and sell. That's because the market works on futures or options. Traders buy and sell the right to buy or sell a given amount of the commodity for delivery at a set time in the future at a guaranteed price. They are betting on the short-term price of the commodities in the hope of profiting from price changes. For example, one trader might agree to buy 5,000 bushels of soybeans at $8 a bushel in 90 days. This assures the soybean growers a certain guaranteed profit, while the trader hopes that the price will rise in the meantime, so that from now until the end of the period he or she can either sell the contract or the soybeans at a higher price. That's why broccoli growers and natural-gas consumers may curse the cold, but those investors who profit from ballooning prices bless it.

The futures market is a kind of insurance against price swings, including those related to weather. This market enables people who actually trade in produce and goods—for example, farmers, sugar refineries, and gas utilities—to reduce the risk they face of losing money because of changes in the price of commodities they haven't yet produced (in the case of the producer) or raw materials they haven't yet received (in the case of the manufacturer). Since the price at which futures in a commodity will be bought and sold depends on the supply and the supply depends, among other things, on weather in the producing area, most people involved in commodities should be keenly interested in what the future weather will be. Veteran traders say the first thing they do in the morning is check the weather forecast at home and overseas.

To gain that competitive advantage, some commodity firms even hire their own private meteorologists to get the detailed weather reports and exclusivity they need. Some meteorologists, like James Roemer of Raleigh, North Carolina, a former television weather broadcaster, have successfully combined forecasting and trading. They tell their newsletter or hot-line subscribers not only what the weather will be but also what to do about it.

Not everyone thinks it's proper for practicing meteorologists to be dabbling in the markets. Something about insider trading! Some meteorologists, though, are so confident of their forecasts that they have parlayed weather hunches into big windfalls. One meteorologist-trader said, "Orange juice and pork bellies put my children through school."

What really practical information can the meteorologist provide the investor-trader? It's not likely to be today's weather forecast—that's available to everyone. What customers really need is precise and detailed, preferably exclusive, information on weather here and elsewhere, what effect current weather will have on the commodity, and what the weather will be a few weeks or few months from now.

In preparing the investor's weather report, the meteorologist sifts through stacks of charts and data from around the world. The forecaster monitors satellite imagery, follows weather systems in the upper atmosphere, monitors surface water temperatures in the tropical oceans, and consults historical records. For those monitoring grains and livestock futures, contemporary weather data are then converted into projections of crop growth and yield potential. That's why records are kept of weather anomalies in the corn belt, the success or failure of the Indian monsoon, and the severity of drought in Australia and Russia.

Monitoring developing and evolving weather conditions such as drought and flooding, and assessing crop health abroad for governments and traders have become necessary activities in order to avoid a repeat of the Great Grain Robbery of 1972. That year, drought and massive crop failures in the Soviet Union led government officials to purchase 20 million tons or nearly 10 percent of the North American grain crop at bargain prices. No one knew how serious the drought and the lack of snow were in the grain-growing territory of the Soviet Union. The subsequent disappearance of grain from the market caused food prices to rise sharply in North America, in spite of a bumper harvest, and a tripling of world cereal prices. The result was regional famine in Africa and South Asia, a scramble for available grain reserves, market speculation, and widespread inflation.

Without question, future weather is what weather intelligence investors most eagerly seek. Brokers and traders know it's much easier to predict tomorrow's Dow Jones average than those for next month. The same holds true for meteorologists trying to forecast into the future. As the time period of the forecast is extended, the level of accuracy drops steadily. It is well known that as soon as a prediction is made beyond seven days, it can only be marginally better than chance.

All long-lead forecasting methods are controversial and not widely accepted. Generally, commodity weather analysts shy away from global circulation models of the atmosphere and oceans that use complex mathematical equations to simulate the atmosphere. They have their own "quick and dirty" methods and don't rely on the national weather service.

Most private meteorologists use analog and statistical methods to forecast future weather. The analog approach is based on the belief that weather patterns repeat. The technique involves matching, say, this autumn with a previous autumn and inferring that perhaps this winter will be similar to that previous year's winter. The trick is to find a previous year that is a perfect match, and that's difficult. True analogs are both rare and poorly behaved.

A statistical method that uses several past years of data is the most common approach to forecasting weather a month or season ahead. Certain abnormalities in the weather are more persistent than others, and certain geographical patterns of weather tend to recur. Trends, cycles, and rhythms do appear in long-term climatic records. The statistical relationships that are found, though, are generally weak, and it is a difficult problem for scientists to determine which of the observed relationships are reliable and which are chance coincidence in the data.

Jon Miller of Smith Barney found that after the 10 warmest winters in 100 years in the United States, eight of the following summers were warmer than normal, with five of the 10 ranking in the top 20 warmest summers of all time. On the other hand, only two of the following winters were warmer than normal, three were close to normal, and five were colder than normal. He also found that extremely warm summers nationwide tend to lead to mild winters across the United States.

Other statistical methods have attempted to correlate weather and sunspot activity. For example, James Roemer found that years immediately following minimum sunspot activity typically were colder than normal in the United States. Persistence is another seasonal forecasting method consulting forecasters use. It implies a continuation of current conditions—that is, what you see is what you're going to get.

**The Blizzard of '96 wasn't all bad for retailers. Whereas severe winter weather causes certain losses, customers often cheer themselves up by making large purchases. In the entertainment and food sectors, the big storm was a boon to video rental stores and take-out vendors, while gambling casinos in Atlantic City—some of which were forced to close—saw a loss of business.**

A very promising area of research in long-range forecasting concerns teleconnections—linkages between weather changes occurring in widely separated regions of the globe often many thousands of miles apart. The best known teleconnection is El Niño.

El Niño seems to have something for everyone. It has been blamed for droughts in Indonesia and Brazil, bush fires in Australia, flooding rains in Chile and Peru, and the failure of the Indian monsoon. In 1997–98, it sent butter prices and romaine lettuce prices through the roof. In persisting for 12 to 18 months, El Niño can seriously affect a country's food supply, economic base, and water resources.

In temperate latitudes over North America, the impacts of El Niño show up most clearly during the winter. For example, most El Niño winters are mild over parts of the northern United States, and wet over California and the southern United States from Texas to Florida. The warm El Niño tends to strengthen the jet stream as it flows from west to east across the Pacific and North America. This enhanced jet stream blocks cold air from penetrating down into the heart of the continent, leaving most of the northern US milder and drier than the normal weather regime. How certain are these temperature and precipitation changes with an El Niño? El Niño effects depend largely on how warm the ocean's surface water gets and how large an area of the Pacific warms up.

Forecasting El Niño's duration and precise effects is difficult. You can't say whether this one will be worse than the previous one. It never stays the same; neither do the impacts. In some cases, though, the direct association between El Niño warming and weather is quite predictable and measurable.

Armed with information about an emerging El Niño, farmers around the world would be better able to assess their risks, and make changes to planting strategies in anticipation of severe farming conditions. Watching El Niño and acting before the weather changes could also have economic benefits for commodities traders.

When El Niño is brewing, meteorologists believe they can make reasonably accurate, credible, and timely weather forecasts for many months ahead and assess the impacts that warming will have on many vulnerable regions around the world. Until other teleconnections are identified and better understood, predicting the weather a month or season ahead is a chancy proposition, given the atmosphere's chaotic behavior.

**Weather derivatives are contracts that obtain their value from the performance of the weather. They are, most simply, a way of protecting your stocks in case of bad weather. For example, in 1998, the Canadian snowmobile manufacturer Bombardier decided, in an effort to boost sales, to offer its customers a rebate of $1,000 if snowfall was less than half of the local average. Bombarier then made a deal with a US energy firm to pay them a small fee for every snowmobile sold. In exchange, the energy company agreed to pick up the tab on the rebates if the snowfall was less than expected. Snowfall was sufficient for snowmobiling that year and no rebates were necessary. While Bombardier had to pay its fee to the US energy company, the deal successfully minimized the company's risks and resulted in a 38 percent increase in sales.**

# Quiz: How Weatherwise Are You?

If you live in North America, you've probably experienced more weather in one year than most people do in an entire lifetime. Consider yourself one of the most weather-astute, weather-conversant, weather-sensitive people around. From coast to coast, the weather, good or bad, is something all North Americans have in common. Also common among North Americans is our fascination with information trivia.

Here's a fun quiz that will show you exactly how absorbed we can get with two of our favorite pastimes: the weather and trivia. Test your weather wisdom by taking this challenging weather quiz. In whole or in part, these fascinating weather-related trivia and factoids are guaranteed to entertain you and should make you a hit at your next social gathering.

Who knows—you may be a genuine weather weenie! Or even better, a supreme weather wizard!

1. The Montreal Expos made Major League history on July 1, 1974, when a game they were hosting at Jarry Park was delayed due to: (a) an earthquake (b) a lightning strike (c) glare from the sun (d) heavy snowflurries (e) baseball-size hail

2. Which of the following golfers has been struck by lightning? (a) Bobby Nichols (b) Lee Trevino (c) Seve Ballesteros (d) Tony Jacklin (e) all of the above

3. Which of the following travels slowest? (a) Brett Hull's slapshot (b) Roger Clemens' fastball (c) a hurricane (d) tornado winds (e) falling hailstones

4. What caused a baseball game indoors at Houston's Astrodome to be postponed on June 15, 1976? (a) rain (b) an earthquake (c) heat and humidity (d) a tsunami warning (e) a bomb threat

5. Which is the least rainy city? (a) New York City (b) Miami (c) Seattle (d) Boston (e) Atlanta

6. If you want to avoid being hit by a hurricane, you should go to: (a) Hawaii (b) New Zealand (c) the equator (d) Maine (e) the North Sea

7. Which country has never launched a weather satellite? (a) Russia (b) China (c) India (d) Japan (e) Canada

8. What is wrong with this statement: last night the mercury fell to −45°F in Buffalo? (a) temperature doesn't really fall (b) Buffalo has never had a temperature that cold (c) with negative temperatures we say temperature rose above (d) below −40°F thermometers use alcohol (e) the National Weather Service advises against going outside to look at thermometers when it's that cold

9. Which continent has never had a hurricane? (a) Asia (b) Africa (c) Australia (d) South America (e) Antarctica

10. Which of the following killer events took the most number of lives, 243? (a) Hurricane Andrew (b) the Mississippi floods in 1994 (c) the East Coast blizzard in March 1993 (d) Hurricane Hugo (e) the eruption of Mount St. Helens

11. The earth is closest to the sun on: (a) September 21 (b) January 3 (c) March 21 (d) June 21 (e) July 3

12. In what 1942 movie did Bing Crosby initially refuse to sing a song that later became his biggest-selling record? (a) *Singin' in the Rain* (b) *Going My Way* (c) *Holiday Inn* (d) *White Christmas* (e) *Stormy Weather*

13. Jet-stream winds are a band of high-altitude winds that circle the globe. They were discovered: (a) when weather satellites were first launched in 1957 (b) by American U2 pilot Gary Powers over the Soviet Union in 1960 (c) by a German Luftwaffe aircraft over the Mediterranean in 1938 (d) at the beginning of the 20th century from weather kites (e) in 1783 when the first hot-air balloon to carry humans ascended

14. The Monroe phenomenon refers to winds that slide down buildings with smooth surfaces and bounce back off the pavement. It is named for: (a) wind engineer Charles Monroe (b) Monroe, Michigan, where it was first detected (c) Marilyn Monroe's problem with her skirts in *The Seven Year Itch* (d) Monroe Towers in Manhattan, New York (e) American president James Monroe

15. What is nephelococcygia? (a) activity of describing the shape of clouds (b) scientific name for weather modification (c) practice of seeding clouds to prevent hail (d) technical name for clouds that resemble flying saucers (e) trance from staring at a full moon

16. What is the average life expectancy of a snow shovel? (a) one season (b) 2.5 years (c) 3 years (d) 5 years (e) 10 years

17. How long does it take a snowflake to reach the ground on its own from a height of two miles? (a) 3 minutes (b) 10 minutes (c) 30 minutes (d) 2 hours (e) 24 hours

18. Which of the following states boasts the largest snowflake—reportedly 15 inches in diameter—ever recorded in the US? (a) Alaska (b) New York (c) Montana (d) North Dakota (e) Maine

19. Of the following locations, which is the safest during a lightning storm? (a) baseball diamond (b) forest (c) swimming pool (d) telephone booth (e) golf course

20. Who said, "The way I see it, if you want the rainbow, you've gotta put up with the rain"? (a) Art Linkletter (b) Monica Lewinsky (c) Boutros Boutros Ghali (d) Dolly Parton (e) The Artist formerly known as Prince

21. The macintosh was invented in 1823 by the Scottish chemist Charles Macintosh. He bound two layers of fabric with a solution of naphtha and Indian rubber. Early macintoshes were called: (a) umbrellas (b) parkas (c) souwesters (d) galoshes (e) raincoats

22. In the open ocean, a tsunami or high-energy wave moves at an incredible speed of up to 620 mph. Ships may not even notice it, yet near shore it can create monster waves that cause enormous damage. What is their principal cause? (a) tidal wave (b) hurricane (c) under-sea earthquake (d) landslide (e) volcano

23. Where would you find the coldest weather? (a) Verkhoyansk, Siberia (b) nine miles above the equator (c) Snag, Yukon (d) the South Pole (e) the thermosphere, about 185 miles above the Earth's surface

24. In which state do most lightning deaths occur? (a) Texas (b) Florida (c) Oklahoma (d) New Jersey (e) Arkansas

25. What trees are the ones most likely to be struck by lightning? (a) fir (b) oak (c) beech (d) pine (e) maple

26. Which of the following locations gets the least rain? (a) the central Sahara (b) Death Valley, California (c) northern Chile (d) Phoenix (e) Antarctica

27. The peak month for tornadoes in the United States is May. On average, Great Britain gets about 40 to 50 tornadoes each year. What is the peak month for British twisters? (a) July (b) June (c) August (d) April (e) October

28. Prior to World War II, meteorologists used to refer to themselves with this nickname: (a) met men (b) rainmakers (c) balloon blowers (d) weather weenies (e) cyclone chasers

29. Under certain weather conditions, at which of these airports is the roar of jetliners so loud during and after takeoff that officials must close the airport? (a) Newark (b) Phoenix (c) Washington (d) Dallas–Fort Worth (e) Chicago

30. Which of the following is a possible name for a future Atlantic hurricane? (a) Hazel (b) Andrew (c) Mitch (d) Pixie (e) Walter

31. In which month did both the hottest and coldest temperatures ever recorded on the surface of the Earth occur? The two temperature extremes spanned almost 270°F. (a) May (b) July (c) September (d) June (e) January

32. Amenomania is a disease that causes victims to become: (a) concerned with the nutritional value of their food (b) morbidly anxious about the direction of the wind (c) lethargic from an excess of positive air ions (d) depressed by long bouts of static weather (e) worried over distant light flickering

33. Operation Desert Storm, the UN-coalition effort to liberate Kuwait, occurred in 1991. What was Operation Typhoon? (a) US air strikes into North Vietnam in 1965 (b) the German attempt to capture Moscow in 1941 (c) a code for Canadian operations in the Pacific front at the close of World War II (d) a US push to re-take Seoul during the Korean War (e) Chinese military maneuvers around Taipei in 1996

34. To South American gauchos, the word *el tormento* refers to what weather term? (a) blizzard (b) sandstorm (c) drizzly fog (d) wind chill (e) dust devil

35. The westerlies is a belt of stormy latitudes. Sailors in the North Atlantic named them: (a) the whirling westerlies (b) horse latitudes (c) Screaming Fifties (d) the trades (e) Roaring Forties

36. An ombrometer and a micropluviometer are technical words for what weather instrument? (a) rain gauge (b) cloud-height sensor (c) satellite cloud camera (d) miniature soil moisture instruments (e) acid snow gauge

37. Recently, green elves have been discovered in the atmosphere. Are they: (a) lightning that occurs high above clouds on the fringes of space (b) air ions that are gradually filling in the ozone hole (c) visible cloud streaks just ahead of a severe thunderstorm (d) contrails from small jetliners (e) plumes spotted by astronauts above the Amazon rain forest

38. What kills more people in North America each year? (a) earthquakes (b) avalanches (c) hang-gliding (d) hurricanes (e) bee stings

39. In meteorology, what does the word *serein* mean? (a) black ice that forms out of the exhaust from tailpipes (b) snow or rain falling from a clear sky (c) a warm comforting breeze (d) a perfect rainbow that occurs from falling snow (e) a green flash at sunset

40. Isobar refers to a line on a weather chart connecting points that have the same value of atmospheric pressure; isotherms refer to equal air pressure. Isobronts are lines of equal: (a) time of some occurrence, such as a wind shift (b) frequency of observing aurora borealis (c) number of hailstorm events (d) amount of thunder (e) depth in water having the same temperature

41. Black rail refers to: (a) slippery streetcar tracks at temperatures near freezing (b) the name of the oldest meteorological station in England (c) ice that forms on roadways from automobile exhaust (d) the blackened center in solar radiation instruments (e) an abandoned railway line

42. What are hypercanes? (a) hurricanes whose maximum wind speed exceeds 155 mph (b) hurricanes that cross the International Date Line changing from hurricane to typhoon or vice versa (c) theoretical superstrong hurricanes whose winds exceed 620 mph (d) hurricanes that kill more than 1,000 people or damage property in excess of $1 billion

43. Potentially, the greatest hazard to scientists chasing severe thunderstorms is: (a) large hail (b) damaging winds (c) a tornado hidden behind rain (d) highway traffic (e) fallen trees and powerlines (f) lightning

44. What US state has the most golfers per capita? (a) Florida (b) California (c) Arizona (d) North Dakota (e) South Carolina

45. Which of the following is not the name of a famous physicist whose name applies to measures of temperature? (a) Fahrenheit (b) Centigrade (c) Reaumur (d) Celsius (e) Kelvin

46. What is a wind rose? (a) an Arctic flower (b) a diagram of proportional directions (c) the name of an instrument to measure wind speed (d) a gentle wind stronger than a calm but lighter than a breeze (e) Beaufort force 2

47. In the US, which month has the greatest average number of tornadoes a year? (a) January (b) February (c) May (d) July (e) August

48. Which state has the highest normal annual temperature? (a) California (b) Texas (c) Florida (d) Louisiana (e) Arizona

49. Which is the driest continent? (a) Asia (b) Africa (c) Australia (d) North America (e) Antarctic

50. Niphablepsia is what form of weather malady? (a) snowblindness (b) frostbite (c) weather migraine (d) prickly heat

51. Yellow snow is: (a) caused by the presence of pine or cypress pollen (b) piled at the end of driveways (c) found only in the mountains of the Sahara Desert (d) synthetic snow

52. Which is the snowiest major city in the US? (a) Boston (b) Syracuse (c) Buffalo (d) Great Falls, Montana (e) Chicago

53. If one considers heat, cold, flood, drought, tornadoes, and tropical storms, which country has the worst weather on Earth? (a) Bangladesh (b) Canada (c) United States (d) China (e) Australia

54. The irrational fear of thunder is termed: (a) keraunophobia (b) katathunphobia (c) cumulonimbusphobia (d) brontophobia (e) sphericphobia

55. What shape are raindrops? (a) teardrops (b) pear shape (c) six-sided crystal (d) round (e) hamburger bun

56. What weather phenomenon kills more people in the developed world than any other natural phenomenon? (a) lightning (b) drought (c) hurricanes (d) tornadoes (e) floods

57. At what temperature does pure alcohol freeze? (a) −38.2°F (b) −113.8°F (c) −148°F (d) −175°F (e) −273°F

58. What place has more rainbows than any other location in the world? (a) Singapore (b) Tampa (c) Glasgow (d) Honolulu (e) Katmandu

59. Which of these weather phenomena are classical music composers most likely to capture in their scores? (a) wind (b) summer heat (c) thunderstorms (d) mist and fog (e) clouds

60. The temperature has fallen below 0°F in every US state except: (a) Florida (b) Louisiana (c) Mississippi (d) Texas (e) Hawaii

61. Which of these cities is the windiest? (a) New York (b) Boston (c) Chicago (d) Phoenix (e) Dallas

62. How many natural disasters in 1998 in the US caused over a billion dollars of damage? (a) 4 (b) 5 (c) 6 (d) 7 (e) 1

63. Of the following cities, which is the sunniest? (a) Phoenix (b) Key West (c) Los Angeles (d) Tampa (e) Fort Lauderdale

64. How many NOAA Weather Radio stations are there in the US and its territories? (a) 20–40 (b) 40–100 (c) 450–500 (d) 600–700 (e) more than 12,000

65. How many meteorological bulletins does the National Weather Service send and receive each day? (a) 50 (b) 500 (c) 40,000 (d) 400,000 (e) 1.4 million

66. How many snowflakes have fallen on the Earth over the past five billion years? (a) equal to the number of cents in the national deficits of all countries in the world (b) 10 million (1 plus 6 zeros) (c) 100 kazillion (indefinite number of zeros) (d) 1 undecillion (1 plus 35 zeros) (e) sexdecillion (1 plus 51 zeros)

67. A zephyr is: (a) the Greek goddess of weather (b) Admiral Beaufort's description of a 20 mph wind (c) a gentle breeze (d) a town in Alaska (e) a snow ruler

68. Which frozen river allowed for the unanticipated victory of George Washington's troops in December 1776? (a) the Hudson (b) the Mississippi (c) the St. Lawrence (d) the Delaware (e) the Allegheny

69. Which President authorized the Secretary of War to establish a national weather service for the US? (a) George Washington (b) Ulysses S. Grant (c) William Taft (d) Herbert Hoover (e) Franklin Roosevelt

70. The US holds the world weather record for which of the following? (a) the most snow in one year (b) the lowest temperature (c) the highest temperature (d) the lowest rainfall (e) none of the above

71. Of the following cities, which has the largest number of cloudy days each year? (a) St. Cloud, Minnesota (b) Seattle, Washington (c) Astoria, Oregon (d) Buffalo, New York (e) Missoula, Montana

72. What is a weather bomb? (a) an erroneous forecast (b) a sudden explosive marine storm (c) a mushroom-shaped cloud (d) a loud clap of thunder without the accompanying lightning (e) a dry tornado

73. Which of the following moves the fastest? (a) snowflakes in a blizzard (b) a hurricane (c) a large hailstone (d) the winner of the Kentucky Derby (e) ice pellets

74. If climate changes occur as climatologists predict, the climate of Vancouver in 2100 will resemble which city today? (a) Los Angeles (b) New York (c) Seattle (d) Juneau (e) San Francisco

75. What is the average lead time for a tornado warning? (a) 12 minutes (b) 8 minutes (c) 5 minutes (d) 1 minute (e) it's impossible to provide lead time for tornado warnings, even with Doppler radar

76. What was the original name of the US National Weather Service? (a) Confederated Weather Service (b) Signal Service of America (c) American Weather Observation Society (d) The Division of Telegrams and Reports for the Benefit of Commerce (e) US Weather Bureau

77. Which of the following is true? (a) pets feel the wind chill (b) you only feel the wind chill under a high-pressure system (c) the wind chill is always lower when the sun is shining (d) in a calm the wind chill is lower than the air temperature

78. Of the following, who did not begin his or her career as a weathercaster? (a) Suzanne Sommers (b) David Letterman (c) Pat Sajak (d) Diane Sawyer (e) Danny De Vito

79. Which cloud is often mistaken for a UFO? (a) altocumulus lenticularis (b) scud (c) roll cloud (d) cumulus congestus (e) cap cloud or pileus cloud

80. Fata Morgana is: (a) a mirage (b) the headquarters of the Italian weather service (c) a tropical storm made famous by having shipwrecked Henry Morgan (d) the inventor of the barometer (e) lightning that strikes the top of a mast

81. Most victims struck and killed by lightning die because of: (a) heart attack (b) third-degree skin burns (c) suffocation (d) exploded intestines (e) charred blood vessels

82. Who invented the alcohol thermometer and the mercury thermometer? (a) Anders Celsius (b) Daniel Fahrenheit (c) William Kelvin (d) Robert Centigrade (e) Marie and Pierre Curie

83. Which name has been called into use most often for North Atlantic hurricanes? (a) Anna/Ana (b) Arlene (c) Debbie/Debby (d) Edith (e) Florence

84. How many weather observation stations were there in the US at the time the Civil War broke out? (a) none (b) 16 (c) 20 (d) 500 (e) 2,000

85. What is the safest place to be during a tornado? (a) bathroom (b) bedroom (c) kitchen (d) front porch or balcony (e) car

86. Which country has the highest participation rate for golf in the world? (a) United States (b) Scotland (c) Switzerland (d) Canada (e) Japan

87. Who said, "Everybody talks about the weather, but nobody does anything about it"? (a) Benjamin Franklin (b) Charles Warner (c) Samuel Clemens (d) Abraham Lincoln (e) Johnny Carson

88. Who is recruited most often as volunteer weather observers? (a) public servants (b) teachers (c) housewives (d) farmers (e) meteorologists

89. What weather phenomenon destroyed crops in France in 1788 and thus contributed to the French Revolution? (a) hail (b) prolonged drought (c) excessive rains (d) frost in the summer (e) lightning

90. Which type of fish can supposedly indicate that bad weather is on the way, by burrowing into mud? (a) goldfish (b) loach (c) salmon (d) marlin (e) arctic char

91. Willy-willies were once hurricanes. Now they are: (a) the Korean name for a blizzard (b) an Australian dust devil or whirlwind (c) a wet bathing suit contest (d) a local Scottish term for cold waves (e) what Californians call snow flurries

92. Which president was also a weather observer? (a) Ben Franklin (b) John Kennedy (c) George Washington (d) Ulysses S. Grant (e) Bill Clinton

93. The US Weather Service costs taxpayers how much per person per year? (a) $1–$2 (b) $5.70 (c) $12.05 (d) $26 (e) over $50

94. When did the word "forecast" come into use in official meteorology? (a) 1776 (b) 1849, when the Smithsonian began to create weather maps (c) 1860, when the Civil War broke out (d) 1869, when a telegraph service in Cincinnati started to collect weather data (e) 1870, when there were only four people working for the Signal Corps

95. When over 400,000 of Napoleon's troops died in the Russian cold, the greatest losses were among those who were: (a) over six feet (b) bald (c) native to the south coast of France (d) front-line soldiers (e) without guns

96. In 1803, Luke Howard, an English weather hobbyist, devised a scheme for classifying cloud types into cirrus, cumulus, and stratus. What was Howard's occupation? (a) Latin professor (b) artist (c) druggist (d) ship captain (e) shepherd

97. What is the average lifetime (in days) of an Atlantic hurricane? (a) 3 (b) 9 (c) 13 (d) 24 (e) 30

98. How many times in its 99-year history has a game in the World Series been postponed because of cold weather? (a) 1 (b) 2 (c) 5 (d) 9 (e) 10

99. According to *Reader's Digest*, the snowiest city in North America is found in: (a) Yukon (b) California (c) Montana (d) Alaska (e) North Dakota

100. Which of the following cities gets the least amount of rain during an average year? (a) Berlin (b) Paris (c) Lisbon (d) London (e) Vienna

# Results

**Scoring 0 to 25**—More of a WEATHER WIMP than a weather weenie. Don't go around criticizing weather forecasters. You need help, but all is not lost! Don't be discouraged—take time to watch the weather, and you'll be surprised what you can learn.

**Scoring 26 to 50**—Good! You're a WEATHER WIT—showing great potential, but there's room for improvement. Just a little more effort will make you a weather weenie. Be sure to watch the skies and listen for interesting weather information.

**Scoring 51 to 75**—Congratulations! You are very well informed on weather and related trivia. Consider yourself a WEATHER WEENIE—it's a great honor!

**Scoring 76 to 100**—Hail, WEATHER WIZARD! You've reached the pinnacle of weather triviadom! You have the right to criticize weather experts.

# Answers

| | | | | | | | |
|---|---|---|---|---|---|---|---|
| 1. | c | 26. | e | 51. | a | 76. | d |
| 2. | e | 27. | e | 52. | b | 77. | a |
| 3. | c | 28. | c | 53. | c | 78. | e |
| 4. | a | 29. | b | 54. | d | 79. | a |
| 5. | c | 30. | e | 55. | e | 80. | a |
| 6. | c | 31. | b | 56. | a | 81. | c |
| 7. | e | 32. | b | 57. | d | 82. | b |
| 8. | d | 33. | b | 58. | d | 83. | b |
| 9. | e | 34. | a | 59. | c | 84. | d |
| 10. | c | 35. | e | 60. | e | 85. | a |
| 11. | b | 36. | a | 61. | b | 86. | d |
| 12. | c | 37. | a | 62. | a | 87. | b |
| 13. | c | 38. | e | 63. | a | 88. | d |
| 14. | c | 39. | b | 64. | c | 89. | a |
| 15. | a | 40. | d | 65. | d | 90. | b |
| 16. | a | 41. | a | 66. | d | 91. | b |
| 17. | c | 42. | c | 67. | c | 92. | c |
| 18. | c | 43. | d | 68. | d | 93. | a |
| 19. | b | 44. | d | 69. | b | 94. | e |
| 20. | d | 45. | b | 70. | e | 95. | b |
| 21. | a | 46. | b | 71. | c | 96. | c |
| 22. | c | 47. | c | 72. | b | 97. | b |
| 23. | b | 48. | c | 73. | c | 98. | a |
| 24. | b | 49. | e | 74. | e | 99. | b |
| 25. | b | 50. | a | 75. | a | 100. | d |

# Glossary

*air mass*: an extensive body of air with a fairly uniform distribution of moisture and temperature throughout

*atmosphere*: the envelope of air surrounding the Earth; most weather events are confined to the lower six miles of the atmosphere

*atmospheric pressure*: the force exerted on the Earth by the weight of the atmosphere

*blizzard*: severe winter weather condition characterized by heavy snow over a sustained period of several hours, low temperatures (less than 20°F), strong winds above 35 miles per hour, and poor visibility (quarter mile or less) due to blowing snow

*blowing snow*: snow lifted from the Earth's surface by the wind to a height of six feet or more; blowing snow rises higher than drifting snow

*bright sunshine*: sunshine intense enough to burn a mark on recording paper mounted in the Campbell-Stokes sunshine recorder; the daily period of bright sunshine is less than that of visible sunshine because the sun's rays are not intense enough to burn the paper just after sunrise, near sunset, and under cloudy conditions

*chinook (snow-eater)*: a dry, warm, strong wind that blows down the eastern slopes of the Rocky Mountains in North America; the warmth and dryness are due principally to heating by compression as the air descends the mountain slope

*climate*: the long-term average that describes what kind of weather can be expected

*crepuscular rays*: rays caused by streaks or beams of sunlight shining through openings in large cumulonimbus clouds on the horizon; the beams reach down and outward from behind the clouds; if they focus upward, towards a point in the sky opposite the sun, they are called anti-crepuscular rays; sometimes they are called sunbeams, crossing the sky, or Jacob's ladder; sailors refer to them as "the sun drawing water"; the dark bands seen crossing the sky are the shadows from clouds

*cyclone*: a generic term that describes all classes of storms from local thunderstorms and tiny dust devils to monstrous hurricanes and typhoons; it comes from the Greek word *kyllon*, meaning "eyele," "circle," or "coil of a snake" and refers to all circular wind systems

*dew-point temperature*: the temperature at which air becomes saturated, allowing condensation of water vapor as frost, fog, dew, mist, or precipitation

*drizzle*: precipitation consisting of numerous minute water droplets that appear to float; the droplets are much smaller than in rain

*El Niño*: Near the end of most years, the normally cold Peru Current, which sweeps northward along the South American coast from southern Chile to the equator, is replaced by a warm southward-flowing coastal current; centuries ago the local fishermen named this the Christ-child current because it appeared around the Christmas season; every few years it was unusually intense and over time the term El Niño became more closely associated with occasional intense warmings

*filling low*: a low in which the central barometric pressure is increasing with time, i.e., the low is gradually weakening

*flash floods*: a very rapid rise of water, most often when an intense thunderstorm drops a huge rainfall on a fairly small area in a very short time

*fog*: a cloud based at the earth's surface consisting of tiny water droplets or, under very cold conditions, ice crystals or ice fog; generally found in calm or low wind conditions; under foggy conditions, visibility is reduced to less than a mile

*frazil ice*: during the freeze-up period, ice forms on a river's surface and ice crystals or frazil develop within the river, especially in open turbulent water slightly below 32°F; frazil ice is very common in rapids

*freezing precipitation*: supercooled water drops of drizzle or rain that freeze on impact to form a coating of ice on the ground or on any objects they strike

*front*: the boundary between two different air masses that have originated from widely separated regions; a cold front is the leading edge of an advancing cold-air mass, and a warm front is the trailing edge of a retreating cold-air mass

*frost*: the deposit of ice crystals that occurs when the air temperature is at or below the freezing point of water; the term frost is also used to describe the icy deposits of water vapor that may form on the ground or on other surfaces like car windshields; these are colder than the surrounding air and have a temperature below freezing

*gale*: a strong wind; a gale warning is issued for expected winds of 40 to 60 mph

*gust*: a sudden, brief increase in wind speed, for generally less than 20 seconds

*heat index*: a measure of what hot weather "feels like"; air of a given temperature and moisture content is equated in comfort to air with a higher temperature and negligible moisture content; at a heat index of 86°F some people begin to experience discomfort

*high pressure*: a term for an area of high (maximum) pressure with a closed, clockwise (in the Northern Hemisphere) circulation of air

*hurricanes*: tropical systems are classed into several categories, depending on maximum strength, usually measured by maximum sustained wind speed; a tropical disturbance is simply a moving area of thunderstorms in the tropics that maintains its identity for 24 hours or more; a tropical depression is a cyclonic system originating over the tropics with a highest sustained wind speed of up to 38 mph; a tropical storm has a highest sustained wind speed of between 39 and 73 mph; a hurricane has wind speeds of 74 mph or more

*ice fog (ice-crystal fog, frozen fog, frost fog, frost flakes, rime fog, pogonip)*: a type of fog composed of suspended particles of ice that occurs at very low temperatures (below −31°F), and usually in clear, calm weather in high latitudes; the sun is usually visible and may cause halos; it is almost always present at air temperatures of −49°F in the vicinity of a source of water vapor, such as fast-flowing streams, herds of animals, and products of combustion for heating or propulsion

*inversion*: a temperature increase with altitude, whereas the usual pattern is a decrease in temperature with altitude

*isobar*: a line on a weather map or chart connecting points of equal pressure; the large concentric lines on television or newspaper weather maps are isobars

*killing frost*: a frost severe enough to end the growing season

*land breeze*: a small-scale wind set off when the air temperature over water is warmer than that over adjacent land; a land breeze develops at night and blows from the land out to the sea or onto a lake; its counterpart is the sea or lake breeze

*low pressure*: an area of low (minimum) atmospheric pressure that has a closed, counterclockwise circulation in the Northern Hemisphere *peak wind (gust)*: the highest instantaneous wind speed recorded for a specific time period

*plough winds*: these belong to a family of strong, straight-line downburst winds found in thunderstorms; they rush to the ground with great force, maybe 60 to 90 mph and occasionally even higher; damage usually covers an area less than two miles across; plough winds are capable of toppling trees, lifting roofs, and ripping apart houses and other structures

*precipitation*: any and all forms of water, whether liquid or solid, that fall from the atmosphere and reach the earth's surface

*probability of precipitation (POP)*: a subjective numerical estimate of the chance of at least .01 inches of precipitation at any given point in the forecast over a specified time period

*relative humidity*: the ratio of water vapor in the air at a given temperature to the maximum that could exist at that temperature; it is usually expressed as a percentage

*ridge*: an elongated area of high pressure extending from the center of a high-pressure region; the opposite of a trough

*sea breeze*: a small-scale wind set off when the air temperature over land is greater than that over the adjacent sea; a sea breeze develops during the day and blows from the sea to the land; its counterpart is the land breeze

*sleet*: frozen raindrops that bounce when they hit the surface; it is not as treacherous to drive on as freezing rain; they have spherical or irregular shapes with a diameter of a fifth of an inch; they do not stick to trees or wires; sleet to a British weather watcher is a mix of rain and partly melted snowflakes

*small-craft warning*: issued when winds over coastal marine areas are expected to reach and maintain speeds of 20 to 33 knots

*snow*: precipitation consisting of white or translucent ice crystals, often agglomerated into snowflakes

*squall*: a strong, sudden wind that generally lasts a few minutes, then quickly decreases in speed; squalls are generally associated with severe thunderstorms

*storm track*: the path taken by a low-pressure center

*storm warning*: the wind warning that is issued to mariners when winds are expected to be 48 to 63 knots

*thunderstorm*: a local storm, usually produced by a cumulonimbus cloud, and always accompanied by thunder and lightning; a thunderstorm day is a day when thunder is heard or when lightning is seen (rain and snow need not have fallen)

*tidal wave*: any unusually high and destructive water level along a shore; also called a storm surge

*tornado*: a violently rotating column of air that is usually visible as a funnel cloud hanging from dark thunderstorm clouds; one of the least extensive of all storms, but in violence the most destructive

*trough*: an elongated area of low pressure extending from the center of a low-pressure region; the opposite of a ridge

*tsunami*: also known (incorrectly) as a tidal wave, *tsunami* is the Japanese word for "harbor wave"; it is a wave set in motion by an undersea movement such as an earthquake or a landslide; these waves can travel up to 600 mph over long distances, hitting the shore with tremendous force

*typhoon*: a severe tropical cyclone in the western Pacific Ocean; the counterpart of the Atlantic hurricane *virga*: streaks of falling rain that evaporate before reaching the ground

*watches and warnings*: the National Weather Service alerts Americans to severe summer storms, winter storms (heavy snow, freezing rain, and blizzards), hurricanes, and floods by issuing weather watches and warnings; usually the first message is the severe thunderstorm watch; if a watch is issued in your area, maintain your normal routine, but keep an eye skyward for threatening weather, and listen to radio and television for further weather information; when severe local storms are building, or have actually been sighted or detected by radar, then warnings are issued and updated; these may be either severe thunderstorm warnings or tornado warnings; warnings mean you should be on the alert

*waterspout*: often called a tornado over water, the actual water spray involved does not extend from the surface to the cloud, but 10 to 30 feet above the water surface; like tornados, waterspouts are very brief; they are capable of overturning small craft or even moving onshore and causing damage

*westerlies (west-wind belt)*: the pronounced west-to-east motion of the atmosphere centered over middle latitudes from about 35 to 65 degrees latitude

*willy-willies*: small, circular winds such as dust devils or whirlwinds in Australia, not very hazardous; before 1950, willy-willies referred to much larger, more destructive typhoons or hurricanes

*wind chill*: a simple measure of the chilling effect experienced by the human body when strong winds are combined with freezing temperatures; the larger the wind chill, the faster the rate of cooling; the wind chill factor is expressed in °F

*wind direction*: the direction from which the wind is blowing

# Index